Life Cycle Analysis Based on Nanoparticles Applied to the Construction Industry

Pilar Mercader-Moyano · Paula Porras-Pereira
Editors

Life Cycle Analysis Based on Nanoparticles Applied to the Construction Industry

A Comprehensive Curriculum

 Springer

Editors
Pilar Mercader-Moyano 
Department of Building Construction I
Higher Technical School of Architecture
University of Seville,
Seville, Spain

Paula Porras-Pereira
Department of Building Construction I
Higher Technical School of Architecture
University of Seville
Seville, Spain

This work was supported by University of Seville, Spain, Business Association for Research of Marble, Stone and Materials, Spain, Delta—Materials process & innovation solutions, Greece, National Technical University of Athens, Greece and Chalmers University of Technology, Sweden.

This Springer imprint is published by the registered company Springer Nature Switzerland AG
The registered company address is: Gewerbestrasse 11, 6330 Cham, Switzerland

If disposing of this product, please recycle the paper.

Preface

In recent years, there has been a notable upsurge in the use of nanomaterials, particularly within the construction and building sectors. The incorporation of nanoparticles introduces noteworthy changes in the physico-mechanical and physical–chemical characteristics of construction materials.

Despite the acknowledged advantages of employing nanomaterials, uncertainties persist regarding their widespread adoption in development and applications, especially concerning potential environmental and human health implications. A crucial aspect in thoroughly assessing the environmental impacts of nanoproducts is the imperative quantification of effects on ecosystems and human health throughout the entire life cycle of these products. The indispensable utilization of a comprehensive tool, such as Life Cycle Assessment, becomes paramount in gaining a nuanced understanding of potential environmental and health challenges, thereby ensuring the environmental sustainability of nanomaterials.

This academic publication endeavours to provide nanoproduct manufacturers, construction industry professionals, and waste managers with the necessary knowledge to comprehend the environmental and health impacts associated with the manufacturing, application, and disposal processes of nanoproducts used in the construction industry. This contribution aims to enhance their personal and professional development, consequently bolstering their employability at the European level.

A thorough comprehension of potential releases throughout the entire life cycle of nanoproducts and their potential effects is imperative for ensuring the safe and sustainable utilization of these innovative materials. The application of life cycle thinking emerges as a pivotal component in appropriately evaluating the potential impacts associated with nanomaterial releases.

Seville, Spain

Pilar Mercader-Moyano
Paula Porras-Pereira

Acknowledgments

The results of this article derive from the project Educational platform for life cycle analysis of treatments based on nanoparticles applied to the construction industry project (code 2022-1-ES01-KA220-HED-000089985), an Erasmus+ project co-funded by the European Union and within the framework of an initiative of 2022 KA220, cooperation partnerships in higher education, with the support of the "Servicio Español para la Internacionalización de la Educación (SEPIE, Spain)".

The European Commission support for the production of this publication does not constitute an endorsement of the contents which reflects the views only of the authors, and the Commission cannot be held responsible for any use which may be made of the information contained therein.

As a final contribution, it is imperative to duly acknowledge the active participation and unwavering commitment of the collaborators associated with this Erasmus+ Project, whose contributions have been instrumental in the realization of the outcomes encapsulated within this book: University of Sevilla, Business Association for Research of Marble, Natural Stone and Materials Technology Center, National Technical University of Athens, Delta Materials Process & Innovation, and Chalmers University of Technology.

Introduction

Buildings exert a direct influence on the environment, extending from the consumption of raw materials in construction, maintenance, and renovation to the discharge of detrimental substances over the building's entire life cycle. The interconnection between the construction industry and the environment is undeniable, with building materials leaving a lasting imprint on both the constructed and natural surroundings throughout their life cycle [1].

In recent years, the utilization of nanomaterials has witnessed a notable upsurge across various applications, particularly within the construction and building sectors. Nanotechnology has proven to be a focal point of uncovering novel applications for nanomaterials, owing to their characteristics such as high-performance attributes, substantial commercial influence, efficacy in energy storage and conversion, cost-effectiveness, energy savings, and diminished environmental impacts. These unique physical, chemical, mechanical, and efficacy characteristics of nanomaterials have sparked a growing interest in their production, surpassing even the use of natural materials, due to the transformative impact of nanomaterials in elevating the performance of building materials [2, 3, 4]. Utilizing nanoparticles introduces noteworthy changes in the physico-mechanical and physical–chemical behaviour of constructional materials. This transformative impact encompasses properties such as increased strength, self-sensing capabilities, self-cleaning features, antimicrobial attributes, and pollution remediation capabilities, all of which can be significantly improved through the incorporation of nanoparticles in construction materials [4].

The significance of these materials became evident as researchers discovered that size could exert a profound influence on the physiochemical properties of a substance [3]. Nanoparticles, owing to their nanoscale dimensions, showcase remarkable attributes, with size being a pivotal factor [4].

Despite the reported benefits of nanomaterial applications for specific industrial applications and consumer products in various sectors, the widespread adoption of nanomaterials in development and applications remain significant uncertainties, particularly concerning their potential environmental and human health impacts if released into the environment [2, 5]. The release of nanomaterials can take place at any stage in the nanomaterial life cycle, spanning from their production, through the manufacturing of nanoproducts, to the utilization/application of these nanoproducts, and finally, their end-of-life management and disposal [5].

To date, much of the research on nanomaterials has centred on their unique functionality in diverse fields and applications, with less attention given to potential environmental effects throughout their life cycle. Concerns also exist regarding the environmental sustainability of nanomaterial pathways contributing to broader environmental problems [2].

The effective incorporation of nanoparticles into construction materials enhances the likelihood of workers being exposed to potentially toxic particles released during construction. Consequently, there is a burgeoning research emphasis on comprehending the toxicity of nanomaterials to the environment and human health, particularly through construction materials, encompassing their manufacturing and recycling processes. It is imperative for humanity to address these issues comprehensively, seeking a thorough understanding of the limitations that nanoparticles may impose on the environment and human health, while also addressing relevant safety concerns. This proactive approach is crucial to ensure responsible and sustainable practices in the utilization of nanomaterials in construction and related industries.

In order to thoroughly evaluate the environmental impacts of nanoproducts, it becomes imperative to quantify the effects on ecosystems and human health arising from releases throughout the entire life cycle of these products [6]. A comprehensive tool, such as Life Cycle Assessment (LCA), proves essential in gaining a better understanding of potential environmental and health challenges and ensuring the environmental sustainability of nanomaterials. LCA, rooted in the concept of life cycle thinking, offers a holistic approach to assessing the environmental impacts of products, services, and processes. It accomplishes this by identifying the materials used and the energy and emissions released to the environment, which is critical in evaluating the potential impacts of nanomaterial releases [2, 5, 7].

A thorough comprehension of potential releases across the entire life cycle and their potential effects is vital for ensuring the safe and sustainable utilization of these new materials. The application of life cycle thinking emerges as a key component in appropriately evaluating the potential impacts associated with nanomaterial releases [5].

After establishing the current significance of nanoparticles in the construction sector, in the context of the Erasmus+ project, a comprehensive series of analyses has been conducted to underscore the significance of integrating Life Cycle Assessment (LCA) into the realm of building materials involving nanoproducts.

The first analysis within this framework entailed the production of a comparative study report designed to systematically gather and scrutinize the existing curricula related to LCA across partner countries, namely Spain, Sweden, and Greece. The primary objective of the comparative study report is to amass information on various instructional resources provided by educational institutions, specifically focusing on training and education in the domain of Life Cycle Assessment.

Moreover, subsequent to a meticulous examination of the respective national reports, the comparative study aspires to accentuate exemplary practices while concurrently identifying any discernible gaps and weaknesses in the current educational landscape pertaining to Life Cycle Assessment. This multifaceted approach aims to contribute to a nuanced understanding of LCA practices and educational strategies, thereby fostering an environment conducive to informed decision-making.

The dearth of training opportunities in Life Cycle Assessments, despite their recognized effectiveness in gauging the environmental impact of products, is conspicuous in the three countries under examination. In an initial exploration, it becomes evident that a significant proportion of the scrutinized academic degrees provide limited exposure to subjects encompassing competences associated with Life Cycle Assessment or Environmental Impact Assessment.

Spain emerges as the nation with the highest inclination towards LCA training, with 44% of the analysed degrees incorporating courses dedicated to LCA-related subjects. In contrast, Greece demonstrates a markedly lower commitment, with only 7% of degrees including such courses, while Sweden falls even below this threshold, with less than 5% of degrees offering LCA-related content. It is noteworthy that the majority of LCA courses are predominantly affiliated with curricula centred on Environmental studies. In the domain of Civil Engineering and Architecture, there is a discernible scarcity of courses dedicated to addressing the intricacies of Life Cycle Assessment.

In addition to the realm of higher education, the findings pertaining to the present curriculum on Life Cycle Assessment and Environmental Impact in the construction sector underscore a substantial gap in the availability of professional courses addressing these crucial subjects. It is noteworthy that within professional education, a significant majority of examined professional associations do not incorporate any courses or programs within their training offerings that encompass competences associated with Life Cycle Assessment or Environmental Impact Assessment. Notably, it is only in Sweden that certain entities provide training specifically geared towards the development of Environmental Product Declarations.

The profound importance of LCA in the construction sector is reflected in its application in various disciplines, since not only Architecture field, but also Civil, Environmental, Building, Industrial, Materials and Architectural Engineering academic areas are focused on Life Cycle Assessment basic principles, methods, and applications as well.

This analysis emphasizes a notable deficiency in educational coverage regarding essential facets of environmental impact assessment. This calls for careful consideration and the possibility of incorporating modules specifically centred on Life Cycle Assessment into pertinent academic programs. It underscores the imperative for a more thorough assimilation of LCA principles into educational frameworks, encompassing both academic and professional realms within the construction industry.

Concluding this comprehensive examination of partner countries' curricula, a parallel study concentrating on nanoparticulate products within the construction sector is concurrently conducted. This study's objective is to gather and assess extant curricula related to nanoparticulate products in the construction sector across partner countries. It involves compiling diverse training guides provided by educational institutions that focus on instruction and education in the realm of nanoparticulate products within the construction sector. Similar to the prior analysis, this comparative study strives to underscore exemplary approaches while discerning any gaps or shortcomings.

Despite the varied applications of nanomaterials in the construction sector and the evident benefits they offer, this analysis reveals a significant deficit in training opportunities in this field. The findings regarding the current curriculum highlight a substantial gap in the provision of courses specifically addressing nanoparticles in the context of construction products, particularly for professionals. Moreover, the results indicate that the majority of courses adopt a combined perspective, encompassing both nanoparticles and construction materials, underscoring the increasing influence of nanotechnology in the construction industry.

Spain emerges as the nation with the highest inclination towards nanoparticles training, with 17% of the analysed degrees incorporating courses dedicated to nanoparticles related subjects. In contrast, Sweden demonstrates a markedly lower commitment, with only 8% of degrees including such courses, while Greece falls even below this threshold, with less than 5% of degrees offering nanoparticles related content.

The primary rationale behind these outcomes is that, despite its unquestionable relevance and innovative nature, the topic of nanoparticles in construction products has yet to be fully developed and expanded across various academic and professional domains.

It is noteworthy that, in Spain, while the previously analysed higher education degrees predominantly lacked subjects related to nanoparticles and their application in the construction sector, professional education tells a different story. In terms of professional training, the only associations providing courses on nanoparticles are those related to architecture. This observation highlights a distinctive trend where architecture-related professional associations seem to be taking the lead in addressing the training gap in nanotechnology within the construction sector.

Despite the current academic landscape, which features a limited percentage of programs offering competences related to nanoparticles or nanoparticulate products in the construction sector, there is a pressing need to broaden the academic scope to encompass more professional sectors and both higher education and professional degrees. This expansion is crucial to fostering innovation and facilitating the adoption of more energy-efficient and sustainable solutions within the construction industry.

Although construction products incorporating nanoparticles are currently scarce, indicating the industry's early stages in this regard, the numerous advantages they present compared to conventional materials suggest a potential significant surge in the use of nanoproducts in the construction sector. This anticipated growth underscores the imminent demand for knowledge pertaining to nanoparticles in construction products.

Higher education plays a pivotal role in preparing a new generation of competent professionals essential for various industries. Therefore, enhancing the incorporation of Life Cycle Assessment into higher education curricula not only catalyses innovation but also contributes to shaping a more sustainable future. Integration of LCA into higher and professional education curricula aids in cultivating a mindset among students and professionals that addresses environmental sustainability through a multidisciplinary approach.

To further promote sustainable and environmentally responsible construction practices and contribute to the overarching goal of establishing a more sustainable and resilient built environment, it becomes imperative to extend the academic offerings related to nanoproducts and LCA to additional professional sectors and both higher education and professional degree programs.

Another of the conducted analyses involved a comparative examination of the normative criteria governing Life Cycle Assessment (LCA) for building materials in Spain, Sweden, and Greece. The principal goal is to discern the commonalities and distinctions in the criteria employed by these countries to regulate the environmental impact associated with building materials.

In striving for this objective, the study seeks to enhance the comprehension of the normative criteria for LCA of building materials and their potential role in advancing sustainable practices within the construction sector. The insights garnered from this investigation have the potential to shape future policies and practices in the construction industry, fostering the adoption of sustainable building materials and practices.

In recent years, numerous countries have implemented regulatory standards for conducting Life Cycle Assessment on building materials, aiming to govern and mitigate their environmental impact. The formulation of these normative criteria exhibits variability influenced by diverse factors, including national policies, regulations, and industry practices. Consequently, undertaking comparative analyses becomes crucial to unravel the nuances and resemblances embedded in the normative criteria for LCA of building materials across distinct countries.

Following the conducted analysis, it is deduced that within each of the examined countries concerning normative criteria for Life Cycle Assessment (LCA) of building materials, there exist three distinct tiers of regulation:

The initial tier of regulation comprises ISO standards, given their international standing. These standards play a pivotal role in the comprehensive framework governing the evaluation of sustainability in construction endeavours. Their primary focus lies in establishing Environmental Product Declarations for construction products, encompassing the formulation of product category rules, communication of life cycle assessment results, and the definition of a coherent and harmonized framework for assessing the environmental impacts of buildings throughout their entire life cycle.

Moving to the second tier, European standards delineate the requirements for Product Category Rules (PCRs) applicable to all construction products and services. These standards furnish the groundwork for conducting Life Cycle Assessments on construction products, spanning all phases of their life cycle.

Concluding the hierarchy at the third level are specific national regulations, such as Royal Decrees, declarations, and certifications. It is noteworthy that a majority of these standards draws upon the international and European standards aforementioned, with minimal discernible variations between them.

As the second and third tiers of regulations draw from ISO standards, they inherently share a common drawback, a generic approach to Life Cycle Assessment that may yield divergent results based on selected parameters. This inherent ambiguity in the normative criteria foundation can introduce variability in the LCA outcomes for building materials. Consequently, there is a critical imperative to prioritize harmonization and standardization in LCA practices, ensuring greater consistency and comparability of results.

This analysis underscores the evident need for continued efforts in unifying and standardizing normative criteria for the LCA of building materials, despite the escalating interest in sustainability and green building practices. Substantial work remains to achieve a more cohesive and consistent framework in this critical domain.

Another of the conducted analyses involved a comprehensive examination of normative criteria related to the application of nanomaterials in Spain, Greece, and Sweden. This examination encompasses technical, occupational health and safety, and environmental protection requirements. The primary objective is to provide the target group with an updated version of all regulations, presented in a more accessible and simplified manner. Additionally, this analysis aims to empower the consortium to develop a nanotechnology application safety training environment in alignment with workplace safety regulations.

In the pursuit of this overarching goal, the study aims to discern the similarities and differences in the criteria employed by these countries to regulate nanoparticles. The insights gleaned from this analysis have the potential to shape future policies and practices within the construction industry. Moreover, they can play a pivotal role in advocating for the adoption of nanoparticulate building materials, thereby contributing to advancements in safety and regulation within the field.

As previously mentioned, the utilization of nanomaterials signifies a revolution in enhancing the performance of construction products. However, notwithstanding the advantages they bring, the safety of workers is facing significant compromise. Nanomaterials pose an invisible threat to workers' health, as their minute size enables interactions at the cellular level. Therefore, a meticulous and proactive approach to addressing safety measures becomes imperative to safeguard the well-being of workers in this evolving landscape of construction technology.

None of the three partner countries has specific regulations addressing the use of nanomaterials in the construction sector. However, amidst this regulatory uncertainty, the European Agency for Safety and Health at Work (EU-OSHA) outlines the existing legislation and underscores the employer's obligation to assess and manage the risks associated with nanomaterials in the workplace, irrespective of the regulatory framework. Currently, these directives, in conjunction with the standards "UNE-EN 482:2012+A1:2016: Exposure in the workplace. General requirements relating to the operation of measurement procedures for chemical agents" and "UNE-EN 689:1996: Atmospheres in the workplace. Guidelines for the assessment of inhalation exposure of chemical agents for comparison with limit values and measurement strategy," form the foundation for ensuring the health and well-being of workers in the construction sector.

In Greece, the regulatory framework applicable to chemical agents, including those used in nanomaterial applications, can be succinctly summarized by the overarching principles outlined in Law 3850/2010 (Government Journal 84/2-6-2010), along with other relevant laws and presidential decrees concerning chemical agents. This regulation emphasizes that employers are obligated to be cognizant of the risks posed to workers' health by factors present or generated in the workplace.

To meet these requirements, employers have the right to solicit information from manufacturers, importers, or suppliers regarding the risks associated with these factors and the safe methods of their use. Furthermore, harmful agents must undergo special treatment, as appropriate, to render them harmless to humans, animals, and the environment before being discharged into the environment. The regulation also specifies that the exposure limit values of chemical substances commonly used in nanomaterial applications have been designated.

In Sweden, the occupational health and safety risks associated with nanoparticles are not a novel concept, and existing laws pertaining to chemicals and dust emissions remain pertinent to the newer nanoparticles introduced in the market, particularly within construction products. The foundation of Swedish legislation lies in assigning responsibilities to employers for the identification of potentially hazardous substances, the assessment of risks linked to these substances, the implementation of relevant mitigation measures, and the verification of the efficacy and utilization of these measures.

This analysis reveals a noticeable legal void, across the partner countries, concerning the advancement of measures aimed at mitigating potential health risks and promoting safe practices in workplaces dealing with chemical agents, particularly those involving nanomaterial applications. This underscores the pressing need for regulatory initiatives and normative actions to address the unique challenges posed by nanoparticles and ensure the protection of workers and the environment in these contexts. Given the acknowledged hazards linked to nanoparticles and the existence of specific EU regulations explicitly addressing them, companies bear a responsibility to undertake appropriate actions. The urgent need for organizations is to adopt a proactive stance in response to these regulatory demands.

In light of the recognized dangers associated with nanoparticles and the specific EU regulations explicitly addressing them, it becomes incumbent upon companies to take necessary measures and to embrace a proactive approach in response to these regulatory imperatives.

Finally, a comprehensive analysis has been conducted on research disseminated in scientific papers. The forthcoming phase encompasses the formulation of a report originating from a comparative examination of existing scientific literature, with a specific emphasis on potential adverse effects on workers' health and the environment arising from the utilization of nanomaterials in the construction sector. The report will also highlight measures aimed at controlling these risks.

This analysis has been divided into two parts: the first involves peer-reviewed environmental studies on nanomaterials, and the second focuses on peer-reviewed studies on the health effects of nanomaterials. Both analyses cover the same types of nanoparticles, those frequently used in the construction sector. For each nanoparticle, its effect on environmentally relevant species and human cells was examined, along with the dose and exposure time necessary to produce a harmful effect.

In most of the analysed cases, the harmful environmental impact of these particles becomes evident after about 24 hours of exposure, whereas for human cells, these effects can be observed with just 4 hours of exposure. These findings highlight the importance of waste management protocols to prevent nanoparticles from being released into the environment. Additionally, they underscore the necessity of comprehensive safety and health protocols that encompass all work environments and protective measures. Informing workers about the risks is crucial, given that negative effects can manifest within just half a workday.

The next stage of this analysis involves the formulation of a report derived from a comparative examination of extant scientific literature that delve into the application of Life Cycle Assessment on building materials, encompassing both qualitative and quantitative dimensions.

In contrast to anticipated outcomes, the majority of studies pertaining to nanoparticles deviate from standardized Life Cycle Assessment methodologies. This deviation poses a challenge to achieving a comprehensive approach for a comparative analysis of their side effects.

A substantial portion of the examined literature demonstrates the utilization of informatic tools in LCA processes. Arguably, the adoption of open-source software is anticipated to enhance data exchange and the establishment of shared knowledge, thereby fostering a greater convergence in results in the future.

While a considerable proportion of the papers delve into at least one mid-point category, fewer than 50% address any end-point category. This outcome aligns with the absence of consensus regarding the assessment of end-point impact categories.

Significant variations exist in the distribution of papers addressing each impact category, indicating a lack of consistency in the applied processes across different studies. Once again, this disparity hampers a comprehensive understanding of the impact of nanoparticles.

The limitations of Life Cycle Assessments on nanoparticles are further compounded by the comparative analysis of system boundaries. Notably, none of the studies encompass the entirety of the cradle-to-cradle cycle, with the majority restricting their analysis to the cradle-to-gate stage, thereby bypassing the in-use stage. This omission is particularly significant when assessing the suitability of nanoparticle-containing materials for construction, as it hinders the evaluation of their side effects during this crucial phase. A comprehensive cradle-to-cradle analysis is imperative to yield a thorough understanding of the risks or damages associated with nanoparticles throughout their entire life cycle.

In future studies within this domain, it is essential to consider a proper and comprehensive functional unit that takes into account all the additional functionalities of nanomaterials. This approach would provide more realistic and equitable assessments of the potential benefits of nanomaterials in advanced technologies. The inadequacy of defining the functional unit introduces higher uncertainty into the study, underscoring the importance of addressing this aspect for more reliable results.

Environmental Product Declarations (EPDs) can serve as valuable sources for datasets related to construction products containing nanoparticles, facilitating their utilization in whole building Life Cycle Assessments and enabling comparisons among different construction products with similar functions. However, it is crucial to exercise caution when employing EPDs, as the comparability of studies may be challenging or even impeded if proper adherence to standards is lacking.

A notable challenge in utilizing EPDs lies in the fact that not all of them encompass the use phase. This is a significant consideration, particularly in the building sector, where nanoparticles can contribute to increased lifespans or enhanced performance, leading to potential savings. Therefore, when assessing construction products that incorporate nanoparticles, it is imperative to consider the inclusion of the use phase in the analysis to obtain a comprehensive understanding of their environmental impact.

The rapid technological progress in nanotechnology necessitates a more in-depth exploration of the environmental toxicity pathways of nanomaterials from a Life Cycle Assessment (LCA) perspective. Understanding the current state-of-the-art LCA applications in nanotechnology is crucial to gaining insights into existing practices and forecasting future developments in this field [2].

The burgeoning growth in the construction sector, fuelled by advancements in nanoscience and technology, is accompanied by an increased utilization of nanomaterials. This surge, however, brings about new complexities in terms of assessment and analysis within the construction industry. The effects of nanoparticles on the environment and human health create substantial gaps in our understanding of their nature and behaviour. In light of research revealing the negative impact of nanomaterials on the environment and human life, it becomes imperative to adopt stringent action plans and guidelines to mitigate environmental risks associated with their use [4].

References

1. G. Ding, Life cycle assessment (LCA) of sustainable building materials: an overview, Life Cycle Assessment (LCA), in *Eco-Labelling and Case Studies* (2014) pp. 38–62
2. N. Nizam, M. Hanafiah, K. Woon, A content review of life cycle assessment of nanomaterials: current practices, challenges, and future prospects, Nanomaterials **11**(12), 3324 (2021)
3. I. Khan, K. Saeed, Nanoparticles: properties, applications and toxicities. Arab. J. Chem. **12**(7), 908–931 (2019)
4. G. Santhosh, G. Nayaka, *Nanoparticles in Construction Industry and Their Toxicity, de Ecological and Health Effects of Building Materials*, ed. by J. Malik, S. Marathe (Springer, 2022)
5. B. Salieri, D. Turner, B. Nowack, R. Hischier, Life cycle assessment of manufactured nanomaterials: where are we? NanoImpact **10**, 108–120 (2018)
6. K. Ettrup, A. Kounina, S. Foss Hansen, J.A.J. Meesters, E.B. Vea, A. Laurent, Correction to development of comparative toxicity potentials of TiO_2 nanoparticles for use in life cycle assessment. Environ. Sci. Tech. **51**(12), 7295 (2017)
7. REACHnano Consortium, Guidance for applying Life Cycle Assessment methodology to nanomaterials, 2015. [Online]. Available: https://invassat.gva.es/documents/161660384/162 311778/02+Guidance+for+applying+Life+Cycle+Assessment+methodology+to+nanomater ials/f40ee359-a279-4464-b2d0-c35c17d5922f. [Accessed: 2023 September 25]

Contents

Materials and Methods

Objectives

The primary objective of the book is contributing to provide nanoproduct manufacturers, construction industry professionals and waste managers with the knowledge necessary to understand the environmental and health impacts of the manufacturing, application, and disposal processes of nanoproducts used in the construction industry. This will contribute to their personal and professional development, thus to their employability at European level.

The publication endeavours to educate and train professionals within the construction sector through a holistic approach on environmental health and safety. Consequently, it is imperative that all agents engaged in various facets of nanoproducts within the construction sector—spanning from nanomaterials and nanoproducts manufacturing, distribution, construction, management, maintenance, and waste management to diverse categories of end-users—actively participate and stay well-informed. Such collective engagement is indispensable for fostering an environment conducive to promoting the health of individuals and the broader environment.

For this purpose, the ensuing specific goals have been established:

- Identification of prevalent nanoproducts in the European construction sector and an exploration of their associated risks.
- Development of a comprehensive curriculum designed to equip individuals engaged with nanoproducts throughout their life cycle with essential knowledge pertaining to environmental impact. This curriculum, aligning with the transformative influence of nanotechnology on traditional industrial processes, aims to train professionals such as chemists, designers, engineers, architects, builders, and waste managers.
- Raising awareness among professionals in the construction sector, manufacturers of nanoproducts, and waste managers regarding the crucial environmental impact of nanoproducts.
- Integration of the curriculum into educational programs to instil knowledge of nanoproducts for upcoming professions.

- Formulation of best practice guidelines for conducting life cycle analyses of nanoproducts.
- Provision of training to targeted groups to enable them to proficiently conduct life cycle analyses of nanoproducts.

Readership

This publication targets entities, organizations, communities, and associations with an interest in comprehending and receiving training regarding the distinct risks associated to nanoproducts in the construction sector, particularly concerning their health and environmental impacts. It is also directed towards research centres engaged in exploring these issues, students, and professionals in the fields of architecture, engineering, chemistry, and waste management, as well as Public Administrations, professionals, and manufacturers actively involved in the construction, maintenance, or management of materials comprising nanoproducts.

Presentation and Structuring of Content

Comprising three parts, this book, while unified by a common thematic context, is designed to be modular, allowing readers to delve into specific sections independently based on individual interests, approaches, or queries. Each chapter is meticulously segmented into distinct sections, a deliberate strategy employed to enhance the overall organization and coherence of the presented information.

The inaugural part, denominated *Fundamental Principles of Sustainability in the Construction Sector: Life Cycle Assessment*, is intricately subdivided into three distinct chapters. The initial segment serves as an introduction, elucidating fundamental concepts regarding the significance of employing Life Cycle Assessment (LCA) for building products incorporating nanoparticles. This chapter encompasses an exploration of the environmental impact generated by the construction sector, an examination of the environment and sustainable development within the framework of Circular Economy principles, and a comprehensive definition of nanoparticles and their contemporary applications. Moving forward, the second chapter systematically acquaints the reader with Life Cycle Assessment (LCA), meticulously defining its primary objectives, scope, and application in the construction sector. This chapter also delineates the normative framework governing Life Cycle Assessment, accompanied by an exploration of its specific terminology. Conclusively, the third chapter meticulously articulates the Life Cycle Assessment methodology, providing a detailed exposition of the various phases inherent in the process, an exploration of diverse impact categories, and an overview of commonly utilized tools for LCA development. Additionally, the chapter sheds light on prevalent environmental certification standards. Furthermore, the chapter culminates with the conclusions elucidating the intricate process of conducting a Life Cycle Assessment for construction products, offering insights into the multifaceted considerations involved in the analysis.

The ensuing part, designated as *Nanoproducts in the Construction Sector*, is structured into three discernible chapters. The initial segment unfolds by expounding upon the foundational concepts of nanotechnology, tracing its historical evolution, delving into its normative framework of reference, and elucidating its diverse applications across various sectors, navigating the inherent challenges associated with nanotechnology. The second chapter is dedicated to an exhaustive elucidation and evaluation of the deployment of nanoproducts in materials integral to the construction sector. This includes a precise definition of the primary nanoproducts prevalent in this sector, accompanied by an in-depth analysis of their most prevalent characterization techniques. Furthermore, the subchapter scrutinizes the environmental and health risks entailed in the application of these nanoproducts within the construction domain. Conclusively, the third chapter aims to provide a comprehensive understanding of the ecological footprint and implications associated with the life cycle of nanoproducts in the construction sector. This chapter unfurls a detailed examination of the potential environmental ramifications stemming from the disposal of nanoproducts, defining waste treatment policies associated with nanoproducts, and conducting an exploration of their life cycle dynamics and an analysis of the pollutant emissions engendered by their application.

The third part, titled *Integration of Life Cycle Assessment and Nanomaterial Assessment*, constitutes the final chapter of the book and meticulously undertakes a comprehensive Life Cycle Assessment of a conventional product, which would be compared with the results obtained from a LCA of a product infused with nanoparticles. Each section within this chapter aligns with distinct phases of the methodology, encompassing the definition of study objectives and scope, development of an inventory detailing consumption and emissions, selection and quantification of environmental impact categories, and the subsequent interpretation of obtained results. The chapter concludes with the synthesis of findings, coupled with formulated recommendations for enhancing the integration of nanoparticles in the construction sector.

The fourth part, entitled *Application of Life Cycle Assessment Results in Building Assessment*, is organized into two distinctive subchapters. The first subchapter delves into the creation of Environmental Product Declarations (EPDs) for construction products incorporating nanoparticles. This involves a comprehensive elucidation of EPDs, detailing the communication of product impact, and providing an overview of prevalent building Life Cycle Assessment (LCA) calculations and tools. The chapter culminates in its second subchapter with an insightful interpretation of LCA results and their comparability, defining the constituent modules and explicating how to analyse diverse scenarios and executing a thorough sensitivity and uncertainty analysis.

Chapter 1
Sustainable Criteria Within the Construction Industry

Juana Esperanza Llorente-García, David Caparrós-Pérez, and María Desiré Alba-Rodríguez

Abstract This initial segment serves as an introduction, elucidating fundamental concepts regarding the significance of employing Life Cycle Assessment (LCA) for building products incorporating nanoparticles. This chapter encompasses an exploration of the environmental impact generated by the construction sector, an examination of the environment and sustainable development within the framework of Circular Economy principles, and a comprehensive definition of nanoparticles and their contemporary applications.

According to the United Nations Environment Programme (UNEP) three planets will be necessary by 2050 to bear the world consumption and our way of life. This will be the case if the current consumption and production is continued, taking into consideration the growing population. Hence, in several fields, there is an increasing focus on sustainability in general and on the environmental impact of different processes in various sectors [1].

Climate change poses one of the most formidable challenges confronting the society today, prompting numerous endeavours aimed at addressing this pressing issue. For instance, as part of the European Climate Change Programme (ECCP) the European Union has set targets for reducing Greenhouse Gas (GHG) emissions in the EU progressively up to 2050 [2]. The latest revisions to these objectives entail

J. E. Llorente-García (✉) · D. Caparrós-Pérez
Department of Sustainable Construction and Architectural Heritage, Technological Centre for Marble, Stone and Materials, Murcia, Spain
e-mail: juana.llorente@ctmarmol.es

D. Caparrós-Pérez
e-mail: david.caparros@ctmarmol.es

M. D. Alba-Rodríguez
Department of Design Engineering, Higher Technical School of Building Engineering, University of Seville, Seville, Spain
e-mail: malba2@us.es

P. Mercader-Moyano and P. Porras-Pereira (eds.), *Life Cycle Analysis Based on Nanoparticles Applied to the Construction Industry*,
https://doi.org/10.1007/978-3-031-79115-4_1

striving to reduce greenhouse gas (GHG) emissions by 55% by the year 2030, with the ultimate goal of attaining climate neutrality by 2050 [3, 4].

1 Environmental Impacts Produced by the Construction Sector

The construction sector, along with its supporting materials industries, stands as a prominent global consumer of natural resources, encompassing both physical and biological realms. Consequently, this industry substantially contributes to the unsustainable trajectory of the global economy [5].

Construction endeavours exert considerable adverse impacts on the environment, manifesting in dust and gas emissions, noise pollution, waste generation, heightened water consumption, and increased air pollution [4]. Notably, in 2021, construction activities regained momentum, returning to levels reminiscent of the pre-pandemic era across the majority of major economies. As a result, buildings energy demand increased by around 4% from 2020—the largest increase in the last 10 years [6].

According to a United Nations report, the construction sector is the largest emitter of greenhouse gases, reaching nearly 38% of global CO_2 emissions [7]. Itemising this Fig. 1, buildings account for around 27% of operational energy related CO_2 emissions, which excludes materials [6]. The emissions from the manufacturing of concrete, steel and aluminium used in the construction of buildings are estimated to represent a further 6% of global emissions [6]. The cement industry is responsible for around 63% of the total CO_2 emissions from the manufacturing process alone [8]. Other materials used in the construction of buildings, such as bricks and glass, are estimated to account for around 2–4% of global emissions. Added together, these would represent around 9% of global operational energy and process-related emissions [6]. According to the report, approximately one third of the world's energy consumption is related to the use of residential and office buildings, and approximately 30–40% of gas emissions come from the construction sector, which, in turn, uses around three billion tons of raw materials, accounting for approximately 40–50% of global extraction levels [8].

Advancing techniques that reduce the emissions from concrete and steel would lead to a fall in embodied emissions for newly constructed buildings, though there is also a strong need to reduce the demand for materials and reuse construction materials more effectively [6].

As stated above, at a worldwide level, construction is responsible for about 40% of carbon emissions, 20–50% of consumption of natural resources, and 50% of total solid waste. A significant percentage of these impacts arises during the operational phase, thereby exerting a substantial environmental footprint [9, 10]. With its vast influence on both the environment and available resources, the construction sector also contributes to environmental degradation, notably through air, soil, and water pollution. Consequently, there is an urgent imperative to mitigate these adverse

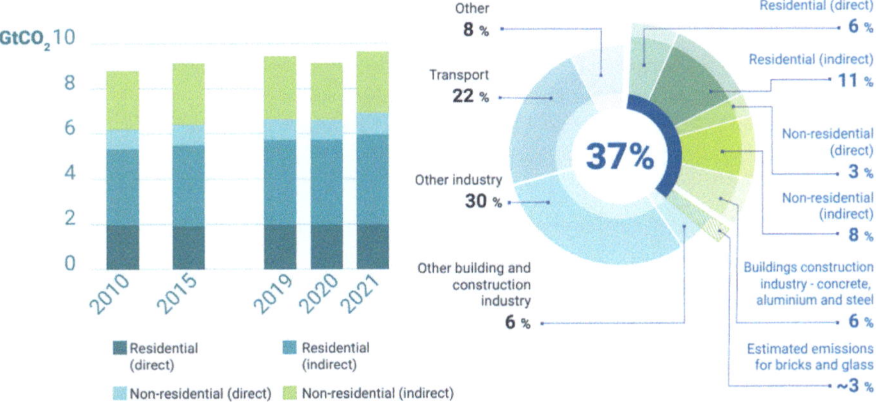

Fig. 1 CO$_2$ emissions in buildings 2010–2021 (left) and share of buildings in global energy and share of buildings in global energy and process emissions in 2021 (right) [6]

effects. It was demonstrated empirically that energy consumption in the construction industry might represent even 40% of the overall energy demand in a developed country [8]. The significance of buildings in addressing climate change issues is paramount, given their substantial global contribution to greenhouse gas emissions and the escalating domestic energy consumption [11].

Acknowledging the building sector's notable role in global carbon dioxide emissions, considerable efforts have been directed towards mitigating this impact [12–14]. A significant portion of the energy consumed within the building sector is attributed to various stages of material production, extraction, high-temperature processes, transportation, and waste management [15].

The construction industry's reliance on extensive quantities of virgin materials, coupled with the cement industry's contribution of 8% to global carbon dioxide emissions, underscores the urgency of addressing these challenges [16]. As the world's population continues to grow, there is a corresponding surge in demand for new materials, exacerbating the strain on natural resources. The prevailing linear economic model perpetuates this pattern, resulting in substantial resource consumption and pollution accumulation throughout the materials' life cycle. These factors present the main driving forces for the endeavour to deal with the negative consequences on the world climate [17].

The construction sector has significant environmental impacts, contributing to various forms of pollution and resource depletion. These impacts can occur at different stages of a construction project, from extraction of raw materials to demolition and disposal of structures [17]. Some of the key environmental impacts produced by the construction sector include:

Resource Depletion:

Raw Material Extraction: The construction industry relies heavily on the extraction of natural resources, such as sand, gravel, timber, and minerals, which can lead to habitat destruction, soil erosion, and depletion of finite resources.

Energy Consumption:

High Energy Usage: Construction activities consume substantial amounts of energy for machinery, transportation, and heating or cooling at construction sites. The energy often comes from fossil fuels, contributing to greenhouse gas emissions.

Waste Generation:

Construction Waste: Construction generates a significant amount of waste, including concrete, wood, metals, plastics, and hazardous materials. Improper disposal can lead to landfills, pollution, and soil contamination.

Demolition Waste: Demolishing buildings also produces substantial waste, which may contain hazardous materials like asbestos or lead.

Air Pollution:

Dust and Particulate Matter: Construction activities generate dust and particulate matter, which can degrade air quality and pose health risks to workers and nearby residents.

Emissions: The operation of heavy machinery and vehicles on construction sites can release air pollutants, including nitrogen oxides (Nox) and volatile organic compounds (VOCs).

Water Pollution:

Runoff and Sedimentation: Construction sites can contribute to water pollution by allowing runoff containing sediment, construction materials, and pollutants to enter nearby water bodies, leading to sedimentation, erosion, and damage to aquatic ecosystems.

Chemical Contamination: Improper storage or disposal of construction chemicals, like paints, solvents, and adhesives, can contaminate water sources.

Habitat Disruption: Construction can disrupt natural habitats, leading to the displacement or extinction of local plant and animal species. Wetlands, forests, and other ecosystems may be destroyed or fragmented.

Noise Pollution: Construction sites are often noisy, and excessive noise can negatively impact the health and well-being of both workers and nearby residents.

Energy-Intensive Building Materials: The production of energy-intensive building materials, such as steel and cement, contributes to greenhouse gas emissions.

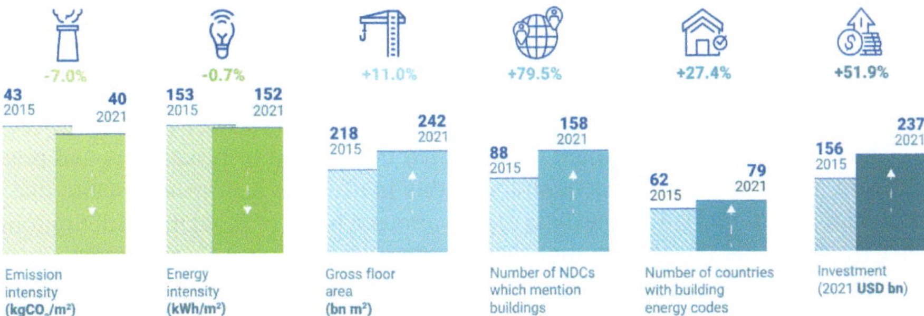

Fig. 2 Global buildings and construction key trends 2015 and 2021 [6]

Urban Heat Island Effect: Construction in urban areas can contribute to the urban heat island effect by replacing green spaces with heat-absorbing surfaces like concrete and asphalt.

Overall, the United Nations "2022 Global Status Report for Buildings and Construction" highlights that since 2015, some progress has been made on the policy level and with an increase in investments, but there must be greater effort to reduce emissions overall and improve building energy performance (Fig. 2) [6].

Nevertheless, the aforementioned data shows that few structural changes have yet occurred within the buildings sector to reduce energy demand or cut emissions. Despite some progress at the policy level, the lack of structural change highlights the growing gap between the actual climate performance of the sector and the necessary decarbonization pathway [6].

The primary issue in sustainable growth with Linear Economy is that it pursues continual economic expansion at the cost of environmental degradation, with no clear grasp of whether this enhances social fairness [18]. As a result, global sustainability concerns are on the rise [19].

2 Environment and Sustainable Development: Circular Economy

"Humanity has the ability to make development sustainable—to ensure that it meets the needs of the present without compromising the ability of future generations to meet their own needs" [20].

Building construction activities have rebounded from the pandemic lows and have been a driver behind both the growing investment in more efficient buildings and the increased global energy demand and related emissions [6]. As aforementioned, the construction industry stands as the foremost consumer of natural resources globally. Historically, this industry has operated under a non-sustainable, linear economic model, adhering to the "take, make, dispose of" paradigm. Regrettably, this approach persists in current practices [21]. The linear approach currently employed does not facilitate the disassembly and repurposing of constructed facilities, rendering them obsolete once they reach the end of their functional lifespan [22]. However, this paradigm must evolve in alignment with the contemporary emphasis on sustainability and worldwide green initiatives [19].

The Circular Economy, which has captured the interest of researchers and practitioners in the last decade, is contrary to this ineffective and unsustainable linear economic paradigm [19]. Regardless of the various schools of thought and definitions, the Circular Economy aims to maintain resources flowing at their best value within boundaries. This guarantees that there is no requirement for new natural resources to manufacture materials and that waste generation is minimized [19]. Aside from resource circularity in closed-loop systems, the Circular Economy focuses on better resource management by rethinking and reducing unnecessary consumption [19]. The Circular Economy is considered a solution to diminish the dependency on resource extraction. It is a condition for preserving the current way of life by sustaining the value of resources and keeping them in circulation [23, 24]. The transition from a Linear Economy to a Circular Economy is a real challenge. However, attaining this objective will contribute significantly to achieving long-term sustainability goals. The Circular Economy is thought to have aided in transforming the conventional Linear Economy into a closed substance economy, which is required and beneficial for developing a sustainable society [19]. The Circular Economy's foundation is built on better resource management through lower consumption and replacing the "end-of-life" concept with the reusing, recycling, and recovering materials and components, a novel approach to achieving sustainable development [19, 25].

The construction industry faces many problems, including a non-linear economy, the absence of financial assistance, lack of proper technology, non-supportive infrastructure, and lack of political will towards sustainable development [19]. The construction industry grapples with complexities arising from various uncertainties, including fluctuating raw material prices, material shortages, rising demand, urbanization trends, impacts of climate change, inadequate waste management infrastructure, and improper recycling technologies [26]. These factors give rise to controversies regarding the root causes of problems and the appropriate strategies for addressing them. However, complexity is beyond the scope of any single organization to comprehend and respond [19]. Implementing Circular Economy principles in construction industry will lower industry costs; reduce negative environmental effects; make urban areas more liveable, productive, and convenient; and help deal with these process complexities [27]. While the concept of the Circular Economy has garnered attention in academia, business, and government circles, its broad adoption and implementation remain limited [28]. The adoption of the Circular Economy concept within the construction industry is still in its early stages. Nevertheless, irrespective of its current level of implementation, the Circular Economy has emerged as one of the most crucial and contemporary approach for addressing sustainability concerns [29].

The United Nations' (UN) 17 Sustainable Development Goals (SDGs) aim to make the world a better place for people and the environment by 2030 [30]. By adopting the SDGs in 2015, 193 countries agreed to address the world's most pressing issues [31]. The Circular Economy contributes to the achievement of the UN's SDGs [32]. This possibility abounds in the many ways in which individuals and corporations engage while enabling the Circular Economy. The Circular Economy can start with the basics and have a big impact, opening new avenues for collaboration to maintain and develop value through revitalized buildings, meaningful occupations, and enhanced mobility [33]. The SDGs provide a new prism through which global needs and objectives can be transformed into economic alternatives for the construction industry. The 17 SDGs can be divided into five categories or the 5Ps: people, planet, prosperity, peace, and partnership [19]. These 5Ps of the SDGs underscore the imperative for collaboration among various stakeholders, such as governments, institutions, and corporations. Given the pivotal role of the construction industry in this framework, it is essential for the industry to devise strategies that align its business strategies with the SDGs. Since the introduction and approval of the UN 2030 agenda, academics and practitioners from various fields have been working hard to research possible SDGs implementation techniques [32]. The construction industry should be no exception to this, and uplift from the lens of UN SDGs is needed while striving to enable Circular Economy in the construction industry. Circular Economy could directly contribute to the achievement of several SDGs, including SDG6 (provide universal access to water and sanitation), SDG7 (affordable and clean energy), SDG8 (inclusive and sustainable economic growth, employment, and decent work), SDG 9 (resilient infrastructure, sustainable industry, and innovation), SDG 11 (make cities inclusive, safe, resilient, and sustainable), SDG12 (responsible consumption and production), and SDG15 (Life on land) [34, 35].

Circular Economy principles are incredibly relevant to the construction industry, with the building sector being a major worldwide consumer of commodities such as energy, materials, and resources [36]. For its high resource intensity, the construction industry draws the most attention throughout the Circular Economy transition [37]. Nevertheless, uncertainties caused by fluctuating raw material prices, shortage of materials, increasing demand, urbanization, climate change, absence of proper waste infrastructure, and use of wrong recycling technologies all lead to complexities in the construction industry [19]. Industries perceive the Circular Economy as a strategy to harmonize economic, societal, and environmental interests by transitioning from linear economies to circular economies, thus maximizing the value of products [38]. Unfortunately, this integration comes at the cost of increased complexity in various ways [39].

3 Waste in the Construction Process

The wreckage developed during the reformation, constructional activities and demolition practices of various structural elements and pavements gives rise to the C&D (construction and demolition) wastes [40]. Based on their generation phase, C&D waste can be classified into three main categories: construction waste (CW), renovation waste (RW), and demolition waste (DW) [41]. Specifically, construction waste encompasses a diverse range of building materials arising from construction activities, which may originate from various stages such as design, procurement, handling, operation, and residual sources. On the other hand, demolition waste comprises materials like wall coverings, paint, paper, aggregate, wood, concrete, fasteners, and adhesives. In broad terms, the components of C&D waste are typically categorized into two groups: major and minor. The former consists of plastic, stone, steel, bricks and wood, whereas the latter contains tiles, paints, glass and electrical fixtures and panels [40]. Typical components in C&D waste are inert materials, which are generally believed have little damage to the environment. Hence, C&D waste is deemed a priority for recycling under the EU Waste Strategy. However, there are also some hazardous components in this particular stream. If these components are not disposed of properly, negative impacts will be made on the environment [41].

In the European Union, according to the amendment to the Act on waste 2014/955/EU: Commission Decision of 18 December 2014 amending Decision 2000/532/EC on the list of waste pursuant to Directive 2008/98/EC of the European Parliament and of the Council [42], waste is "a substance and/or an object which the holder concerned intends or is required to discard."[42–44]. Waste management encompasses the collection, transportation, and processing of waste, including sorting activities, along with the oversight of these processes. Additionally, it involves the subsequent management of waste disposal sites and activities carried out as a waste seller or waste broker [45]. On the contrary, according to the aforementioned Act, waste management is "waste generation and management" [45], and material recovery is understood as "any recovery other than energy recovery and reprocessing into materials that can be used as fuels or other means of energy production; material recovery includes, preparation for re-use, recycling and earthworks" [8, 45].

Many current construction materials rely on energy intensive, mineral-based extractive processes which cause deleterious environmental impacts across the material life cycle, such as biodiversity loss and water scarcity as well as contributing to both embodied and operational carbon emissions. Additionally, at the end-of-use phase of building systems and infrastructure, materials are often wasted, exacerbating the environmental impacts associated with current 'take-make-waste' linear material production practices. Globally, approximately 100 billion tonnes of waste is caused by construction, renovation and demolition, with about 35% sent to landfills whereas it could be recovered and upcycled. In fast-growing developing economies, construction materials are set to dominate resource consumption, with associated greenhouse gas emissions expected to double by 2060 [6]. In the European Union, these construction and demolition projects are also responsible for about a third of the total waste generated, with a significant share being landfilled which creates serious environmental problems during the entire lifecycle of buildings, especially during the operation and end-of-life stages [46]. The C&D waste management has been a grave matter worldwide owing to the significant amount of C&D waste generation across the globe (Fig. 3) [40].

The categorization of waste is contingent upon its origins and the level of hazard or potential harm it poses to health, human life, animals, or the environment. Waste can be classified into various categories based on factors such as its source (consumption

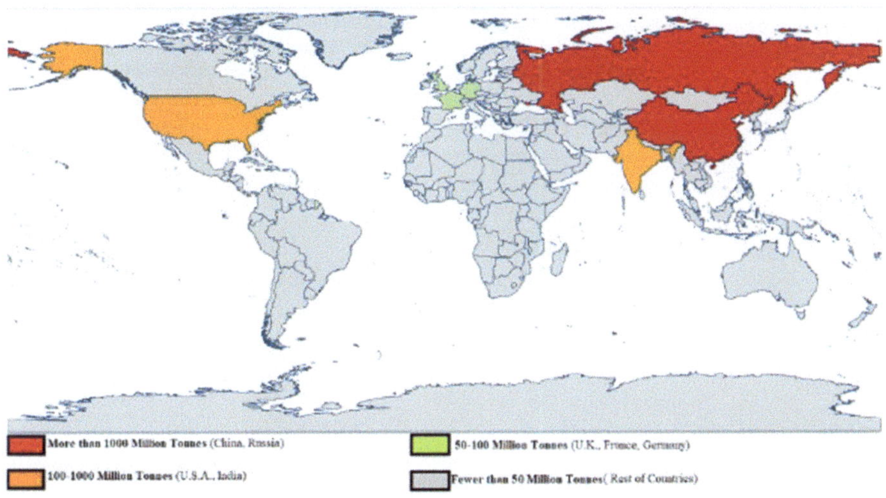

Fig. 3   C&D waste created crosswise the globe [40]

or production), degree of hazard, properties, and origin [45]. Within the European Union, waste is classified into 20 groups based on their respective origins [45]. In recent years, particularly within the construction sector, notably the housing industry, a noticeable trend towards "low environmental impact" has emerged. This trend emphasizes the utilization of materials sourced from natural origins, as well as waste and recycled materials. Such solutions are seen as cost-effective, energy-efficient, and beneficial for both health and the environment. A significant advantage of these approaches lies in the utilization of waste materials [8, 45].

According to a 2019 OECD report, the global consumption of raw materials will almost double by 2060 as the world economy grows and living standards rise, exacerbating the environmental overloading we are experiencing today [6]. Based on the report, the most significant rise in resource consumption by 2060 is projected to occur in minerals, encompassing construction materials and metals, especially in rapidly developing economies [6]. The extensive extraction of natural resources, coupled with inefficient utilization of end-of-life materials, leads to the generation of substantial volumes of waste. There is a need to reassess the existing material streams in order to alleviate environmental pressures and effectively attain sustainability objectives [17].

In G7 countries, material efficiency strategies, including the use of recycled materials, could reduce greenhouse gas emissions in the material cycle of residential buildings by 80–100 per cent in 2050 [6]. Comprehensive measures must be implemented across the sector, adopting a holistic and systems-thinking approach that considers the entire life cycle. This approach would facilitate engagement with multiple stakeholders and encourage collaboration across various industries (Fig. 4) [6].

Fig. 4 Whole-life and systems-thinking approach to enable multiple stakeholders at each decision point [6]

Transitioning to a more sustainable future requires the design of multi-beneficial material strategies that take a whole building life cycle and systems-thinking approach. Nevertheless, the foremost priority should be to enhance the longevity of both existing and new building stock, while also maximizing the reuse of existing components whenever possible. A whole life cycle analysis approach is increasingly being adopted by industry leaders to guide strategies to address this problematic. These strategies can be categorized into three main approaches: "avoid," "shift," and "improve", all of which ultimately contribute to enhance "adaptability". The measures encompass a spectrum, from reducing construction, minimizing material requirements, and utilizing low-carbon materials, to adopting circular approaches and enhancing designs with longer lifespans and reduced operational emissions during building use (Fig. 5) [6].

Due to environmental imperatives, there is a critical need for large-scale recycling of C&D waste [47]. Utilizing C&D leftovers aids in reducing the demand for new resources, conserving land for future urbanization, safeguarding the environment and ecology, reducing transportation and energy production costs, and diverting waste from landfills [40]. Integrating C&D debris as a substitute for natural aggregates helps mitigate the ecological footprint associated with the construction sector [40]. This substitution proves valuable, potentially saving up to 60% of natural aggregates, as indicated in a study [47]. Given the scarcity of natural resources, repurposing C&D waste proves instrumental in addressing the heightened demand for aggregates within the construction sector [40].

As a result, the establishment of an efficient C&D waste management system, which is both environmentally sustainable and economically viable, has become a pressing global issue necessitating in-depth exploration and discussion [41]. Recognizing the significance of accurately quantifying C&D waste, it is imperative at both project and regional levels. At the project level, quantification involves forecasting C&D waste production for a specific project. This enables project managers to adjust material purchase schedules, organize on-site stockpiling, and assess the

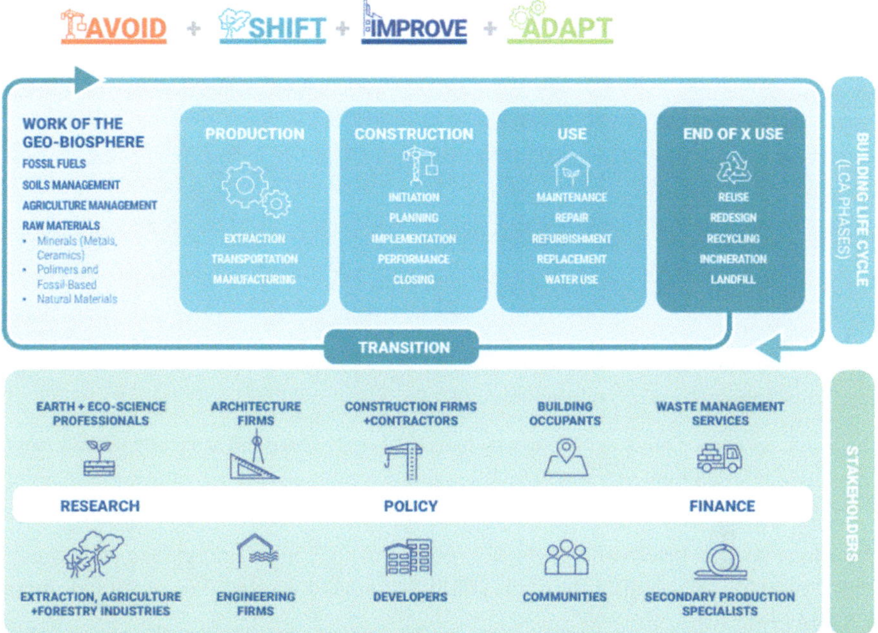

Fig. 5 How to ensure stakeholders receive pertinent information at each phase of the built environment process life cycle to facilitate maximum decarbonization through a systems-thinking approach [6]

potential benefits of waste recycling and disposal costs. On a regional scale, quantification entails estimating the total C&D waste generation across all projects within a designated area [41].

4 Energy in the Process of Sustainable Construction

Operational energy demand in buildings (for space heating and cooling, water heating, lighting, cooking and other uses) accounts for around 30% of final demand, which is an increase of around 4% from 2020 and exceeds the previous peak in 2019 by 3%. The change in energy demand in buildings in 2021 reflects a complex picture. On the one hand, buildings are being operated more intensively than during the pandemic as workplaces reopen and businesses resume more "normal" operations. Equally many workplaces are choosing to maintain hybrid working policies, so a distribution of energy demand in both workplaces and homes remains. Energy used for buildings to produce concrete, steel, and aluminium accounts for a further 4% of final energy demand [6]. Together with operational energy demand this brings buildings energy demand to 34% (Fig. 6). Other materials used in the construction of

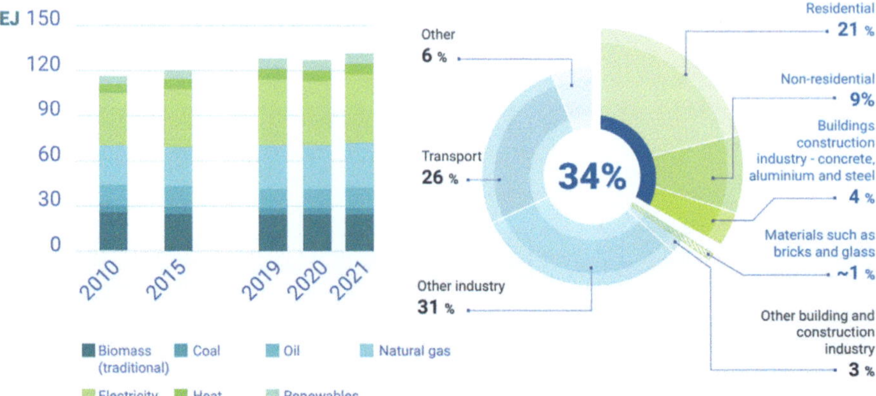

Fig. 6 Energy consumption in buildings by fuel, 2010–2021 (left), and share of buildings in total final energy consumptions in 2021 (right) [6]

buildings, such as bricks and glass, also contribute to global energy use and together with concrete, steel and aluminium could account to around 5% of global energy demand [48].

Similarly, in low- and middle-income countries, access to modern fuels was disrupted due to the pandemic, as highlighted in the latest SDG 7 progress report. This report indicates that 733 million people lack access to electricity, while approximately 2.4 billion people rely on polluting fuels for cooking, adversely affecting their health. The countries with the largest number of people without access to electricity were within Central and Western Africa, with 442 million (6%) being within the African continent. These levels of energy access have improved from 2010 but are off track to meeting the commitments of the Sustainable Development goals by 2030 [6].

The International Energy Agency (IEA) estimates that, in 2021, around 8% of operational energy and process-related CO_2 emissions were from the direct use of fossil fuels in buildings (i.e., direct emissions) and a further 19% were due to electricity use (i.e., indirect emissions). The IEA ascribes the increase in direct CO_2 emissions to increased use of fossil fuels in both advanced and emerging economies, particularly fossil fuel gas in emerging economies [6].

5　The Impact of Transport on Sustainable Criteria

Transportation infrastructure, serving as an intricate network, interlinks cities and facilitates human activities, thereby integrating social, economic, and environmental systems amid urbanization and population expansion. Moreover, this network plays a pivotal role in advancing socioeconomic development and enhancing overall quality of life by fostering inter- and intra-city connectivity during urbanization [49–51].

Furthermore, it is imperative to prioritize objectives such as low-carbon emissions, resilience, and sustainable development when expanding the transportation network [52].

Transportation carries both benefits and drawbacks for society. On the one hand, mobility, or the ability to move, is a basic human need that enables social participation and leads to urban aggregation and diffusion, greatly boosting the regional and national economic development [51]. In 2016, transportation, created 5% of the gross value added in the European Union [50]. On the other hand, growing traffic volumes have led the transportation sector to contribute significantly to ecological destruction, climate change and overall greenhouse gases (GHGs) [51, 53].

Transport is responsible for a quarter of the EU's greenhouse gas emissions, with road transport representing the greatest share, 72% in 2019 [53]. In 2020, the transport sector accounted for 57% of global oil demand and 28% of total energy consumption. Figure 7 shows the development of CO_2 emissions for different transport modes for the 2000–2018 time period, with passenger road vehicle transport being by far the largest contributor. In contrast, rail accounted for only about 0.3% of direct CO_2 emissions [54].

As seen in Fig. 8, between 2010 and 2019, transport CO_2 emissions rose in all regions, except Europe, where they fell 2%, a drop attributed to advanced fuel economy regulations and advancing initiatives on sustainable urban mobility [54].

Climate and energy policies within the EU have led to substantial reductions in greenhouse gas emissions across all sectors, with the exception of transport. Between 1990 and 2019, total transport greenhouse gas emissions increased by over 33%, and road transport emissions surged by almost 28% [53, 55]. Despite existing policy measures, the European Commission projects that transport carbon dioxide (CO_2) emissions will be 3.5% higher in 2030 than in 1990 and are expected to decrease by only 22% by 2050 compared to 1990 levels (Fig. 9). This is a long way from the 90% reduction by 2050 that is needed from transport to achieve the overall 2050 climate neutrality target [55]. While emissions from road transport are projected to fall over time, it will remain the most important sector in terms of its contribution [56]. This highlights the substantial contribution of society's predominant reliance on cars to high greenhouse gas (GHG) emission rates. Given that GHGs are the primary drivers

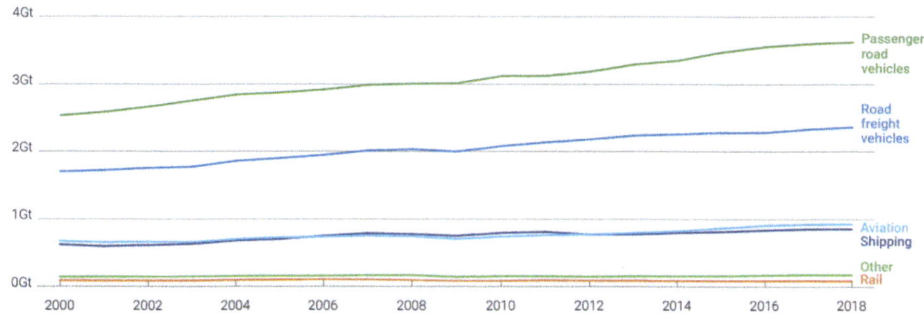

Fig. 7 Transport sector CO2 emissions by mode (2000–2018) [54]

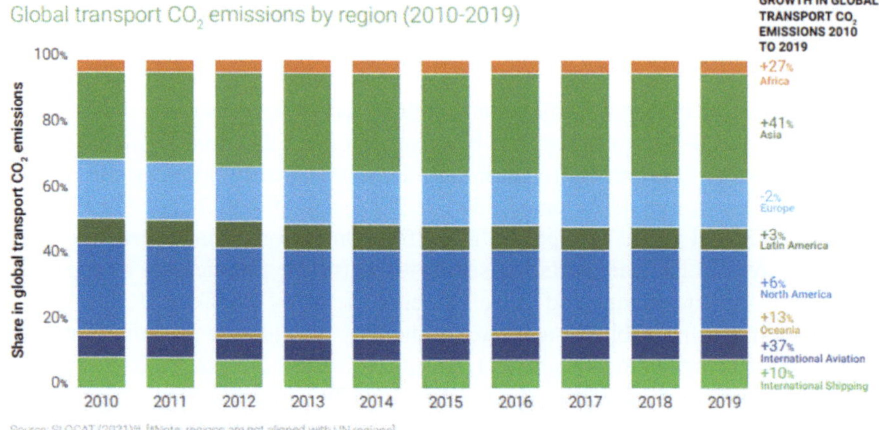

Fig. 8 Global transport CO_2 emissions by region (2010–2019) [54]

of global warming, transportation emerges as a pivotal factor in achieving the Paris Agreement goal of limiting the global temperature rise [53].

Further direct and indirect negative effects caused by the current transportation system include an increase in inequality regarding accessibility, congestion (which harms public health), unsustainable resource use, and accidents, among others [53].

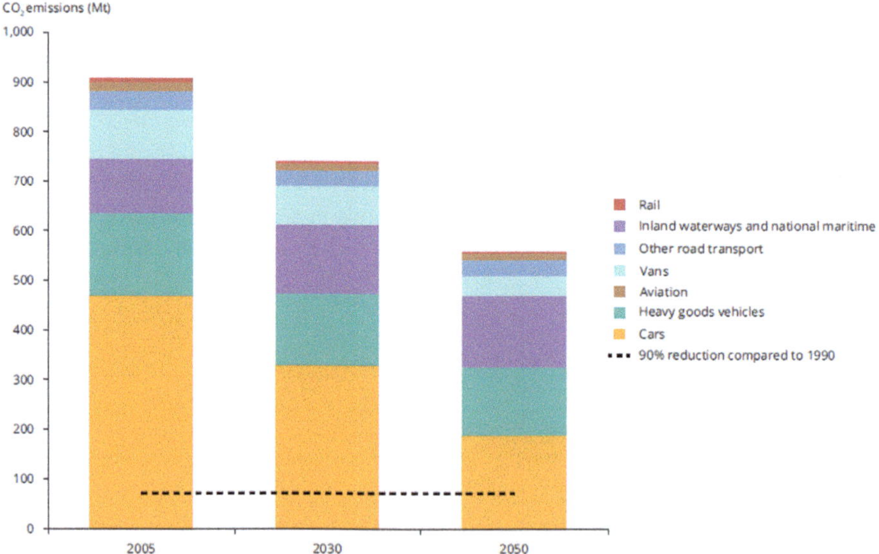

Fig. 9 CO_2 emissions from the transport sector (including international aviation but excluding international maritime) in the EU Reference Scenario 2020 [54]

Thus, there is an urgent need for transformative action that will accelerate the transition to sustainable transport at the global level, since progress to date with regard to sustainable transport has been insufficient. Flows of people, goods, and information have surged significantly in recent decades, with projections indicating continued growth alongside increased demand for sustainable transport [54]. Promoting sustainable transportation concepts is paramount to sustaining mobility while effectively managing transportation in an environmentally sound, socially acceptable, and economically efficient manner. Sustainable transportation concepts are indispensable for mitigating global warming and fostering social well-being [53].

Transportation, as a core component supporting the interactions and development of socioeconomic systems, has been the object of much consideration as to what extent it is sustainable. Sustainable transportation refers to the ability to fulfil the mobility requirements of a society in a manner that minimizes harm to the environment and ensures that the mobility needs of future generations are not compromised [49].

Sustainable transport encompasses the infrastructure that enables individuals and societies to meet their fundamental access needs safely and in a manner that aligns with human and ecosystem health, as well as equity within and between generations. It aims to provide safe, affordable, accessible, efficient, and resilient transportation solutions. Moreover, a sustainable transportation system minimizes emissions, environmental impact, consumption of non-renewable resources, wasting, and the use of land and production of noise (Fig. 10). Therefore, sustainable transport is not an end in itself but rather a tool to achieve sustainable development [49, 57].

Sustainable transport plays a pivotal role in sustainable development, intersecting with various Sustainable Development Goals (SDGs) and serving as a cornerstone

Fig. 10 The three essential factors of transportation system sustainability [57]

for achieving the 2030 Agenda for sustainable development and the Paris Agreement on Climate Change. Beyond facilitating the mobility of people and goods through services and infrastructure, sustainable transport acts as a multifaceted catalyst, accelerating progress towards critical objectives such as poverty eradication, inequality reduction, gender empowerment, improved access to essential services, and climate change mitigation. Furthermore, it contributes to the realization of human rights. Hence, sustainable transport is indispensable for advancing the 2030 Agenda for Sustainable Development and the Paris Climate Change Agreement [54].

Applying sustainable development principles to transport systems entails fostering connections between environmental protection, economic efficiency, and social progress. Anticipated outcomes of sustainable transport encompass enhancements in efficiency, safety, and environmental preservation. Within the environmental dimension, the goal is to comprehend the reciprocal influences between the physical environment and industry practices, ensuring that all facets of the transport sector address environmental concerns. Economically, the aim is to promote efficiency by encouraging the provision of necessary infrastructure and mobility systems. Transport should be cost-effective and adaptable to evolving demands. Socially, the objective is to elevate living standards and improve quality of life [49]. Accelerating the transition to sustainable transport requires intensified national efforts, multi-stakeholder partnerships, and international cooperation [54].

Scientific advancements and the swift adoption of new technologies are imperative for achieving the transition to sustainable transport at the necessary scale and pace. Embedded safety mechanisms, eco-friendly fuels and engines, pervasive digitalization, real-time information processing applications, autonomous vehicles, and intelligent transport systems are now core elements of the transport innovation arena. Ongoing research is vital not only for refining existing solutions but also for developing novel ones to address more challenging issues, such as emissions that are difficult to mitigate from specific transport modes [54].

Environmental sustainability has emerged as an increasingly focal point for transportation service providers, prompting them to prioritize the acquisition of expertise in environmental management. The primary challenge lies in integrating environmentally sustainable transport practices within competitive market frameworks, while also addressing changes in transportation demand and enhancing transport supply [58]. One of the aims of the European Union's transport policy is to achieve that at least 50% of freight is carried by rail and water, for more than 300 kms by 2040. Countries implementing this policy are in the process of establishing public logistics centres in most strategically suitable locations of the country [1]. This is to ensure the sustainable development of transport [59].

Hence, it can be concluded that, although there are positive developments in the transportation sector, more efforts are required to address the fragmentation within the transportation sector and facilitate cohesive, coordinated efforts among the transport community toward achieving a sustainable transformation in transport that ensures inclusivity and leaves no one behind [54]. The need for a set of universally applicable criteria to assess the sustainability of holistic urban transportation is unquestionable [53].

As it has been demonstrated before, transportation plays a key role in minimizing the impact of climate change. It is, therefore, of the utmost importance to make transportation more sustainable, but also, socially acceptable, and economically efficient [53].

6 Definition of Nanoparticles. Current Application

In recent years, nanotechnology is one of the most famous concepts due to the great development of the world of science and technology. Nanotechnology has been a recognized field of research since the last century. Since Nobel laureate Richard P. Feynman introduced nanotechnology in his renowned 1959 lecture "There's Plenty of Room at the Bottom," the field has witnessed numerous ground breaking advancements [60].

Nanotechnology could be defined as the discipline focused on the study, design, synthesis, manipulation and application of materials, devices and functional systems, through the control of matter at the nanoscale and the exploitation of phenomena and properties of matter at the nanoscale [61].

Nanotechnology has facilitated the production of materials across various types at the nanoscale level. Nanoparticles (NPs) are wide class of materials that include articulate substances, which have one dimension less than 100 nm at least. The significance of these materials became apparent when researchers discovered that size can significantly impact the physiochemical properties of a substance, such as its optical properties [60].

Nanoparticles are complex entities, consisting of three distinct layers: (a) The surface layer, which can be functionalized with various small molecules, metal ions, surfactants, and polymers, (b) the shell layer, a chemically distinct material from the core in all aspects, and (c) the core, which is essentially the central portion of the NP and usually refers the NP itself [62].

The US Environmental Agency has classified nanomaterials into four types according to their main component: carbon-based nanomaterials, metal-based nanomaterials, dendrimers, and composite nanomaterials. Carbon-based ones with an ellipsoidal or spherical shape are known as fullerenes, while cylindrical ones are called nanotubes. Metal-based ones include quantum dots, gold and silver nanoparticles, and metal oxides such as titanium dioxide. Dendrimers are nanoscale polymers constructed from branched units, featuring numerous terminal groups on their surface and internal cavities capable of encapsulating other molecules. Composite nanomaterials combine different types of nanoparticles or nanoparticles with larger materials [61].

Dendrimers are nanometric-sized polymers constructed from branched units, a surface with numerous chains ends and interior cavities into which other molecules, such as drugs, can be introduced. Compounds combine nanoparticles with other nanoparticles or with larger materials [61].

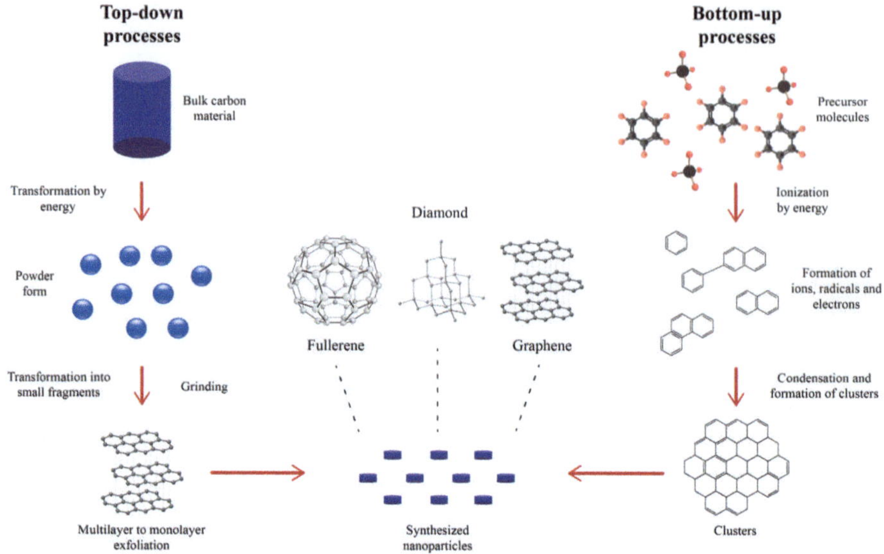

Fig. 11 Nanoparticle synthesis techniques [41]

Furthermore, according to the origin of nanomaterials, they are classified as natural, being produced by trees, plants, volcanoes or marine species; incidental, when they arise during combustion in vehicles and industrial processes; and artificial, the most common one, produced by two manufacturing processes: top-down/bottom-up (Fig. 11). On one hand, top-down techniques consist of the division of macroscopic material or group of solid materials until reaching the nanometric size. Physical methods such as grinding or wear, chemical methods and the volatilization of a solid followed by the condensation of the volatilized components are used, until a series of assemblies are obtained that are precisely controlled until reaching the desired size. On the other hand, bottom-up techniques consist of the manufacture of nanoparticles with the capacity to self-assemble or self-organize through the condensation of atoms or molecular entities in a gas phase or in solution [61].

7 The Impact of Nanoparticle Composite Materials on Sustainable Construction

Nanoparticles are produced from different types of metals such as gold, iron, platinum or metal oxides. Currently, the most used and characterized nanoparticles are those synthesized from silver ions (AgNPs), due to their physical (conductivity), chemical (stability) and biological (catalytic and antibacterial activity) properties [61].

Many of the adverse effects of nanoparticle synthesis have been associated with the toxicity of physical and chemical methods, due to the presence of toxic substances absorbed on the surface of the NP [58, 61].

Eco-friendly alternatives are biological methods for the synthesis of nanoparticles using microorganisms, enzymes, fungi, and plant extracts. The development of these ecosystem-friendly methods for nanoparticle synthesis has become an important branch of nanotechnology: green synthesis [61, 63].

Traditional nanoparticle production uses toxic materials such as solvents and surfactants that can affect the environment. Green synthesis is an alternative technique for bioproduction of nanoparticulate material together with metallic materials (gold, silver, iron, and metal oxides), which seeks to be environmentally friendly [64].

Green synthesis is based on the reduction of metals using natural species with antioxidant power. During the last decade, it has been shown that many biological systems can replace chemical reducing agents including plants and algae, diatoms, bacteria, yeasts, fungi, viruses and human cells [61].

Microorganisms are able to transform inorganic metal ions into metal nanoparticles through the reducing capacities of their metabolites and proteins. The synthesis of nanoparticles can be carried out at an intra- or extracellular level, this is how plants are able to reduce the same inorganic ions into metal nanoparticles both on their plant surface and in certain tissues [63].

Different strategies with plants have been developed for the synthesis of nanoparticles, some use metal salts (example: silver nitrate) or metal oxides (example: titanium oxide) during plant growth and then the nanoparticles are extracted from the dry material of plants [61].

Another way to directly synthesize nanoparticles is through the use of aqueous plant extracts, which contain one or more active ingredients from a specific plant. The use of plant extracts to synthesize nanoparticles is the fastest method [61].

References.

1. C. Dossche, V. Boel, W. De Corte, Use of life cycle assessments in the construction sector: critical review. Procedia Eng. **171**, 302–311 (2017)
2. European Commission, in *Climate Action: Progress Report*, Dec 2022. [Online]. Available: https://climate.ec.europa.eu/system/files/2022-12/com_2022_514_web_en.pdf. Accessed 03 Sept 2023
3. European Commission, in *Stepping up Europe's 2030 Climate Ambition Investing in a Climate-Neutral Future for the Benefit of Our People*, Sept 2020. [Online]. Available: https://eur-lex.europa.eu/legal-content/EN/TXT/?uri=CELEX%3A52020DC0562. Accessed 03 Sept 2023
4. P. Dräger, P. Letmathe, Value losses and environmental impacts in the construction industry—tradeoffs or correlates? J. Clean. Prod. **336**, 130435 (2022)
5. R. Spence, H. Mulligan, Sustainable development and the construction industry. Habitat Int. **19**(3), 279–292 (1995)
6. UN, in *2022 Global Status Report for Buildings and Construction*, Feb 2023. [Online]. Available: https://globalabc.org/sites/default/files/2023-03/2022%20Global%20Status%20Report%20for%20Buildings%20and%20Construction_1.pdf. Accessed 08 Sept 2023

7. IPCC Report, in *Climate Change 2021: The Physical Science Basis, the Working Group I Contribution to the Sixth Assessment Report,* July 2022. [Online]. Available: https://www.ipcc.ch/report/sixth-assessment-report-working-group-i/. Accessed 08 Sept 2023

8. I. Ryłko-Polak, W. Komala, A. Białowiec, The reuse of biomass and industrial waste in biocomposite construction materials for decreasing natural resource use and mitigating the environmental impact of the construction industry: a review. Materials **15**(12), 4078 (2022)

9. M. Khasreen, P. Banfill, G. Menzies, Life-cycle assessment and the environmental impact of buildings: a review. Sustainability **1**(3), 674–701 (2009)

10. A.J. Probert, A. Miller, K. Ip, K.P. Beckett, R. Schofield, in *Accounting for the Life-Cycle Carbon Emissions of New Dwellings in the U.K.* 12th International Conference on Non-Conventional Materials and Technologies, (2010) pp 1–13

11. I. Vasilca, M. Nen, O. Chivu, V. Radu, C. Simion, N. Marinescu, The management of environmental resources in the construction sector: an empirical model. Energies **14**(9), 2489 (2021)

12. E. Açıkkalp, A. Hepbasli, C. Yucer, T. Karakoc, Advanced life cycle integrated exergoeconomic analysis of building heating systems: an application and proposing new indices. J. Clean. Prod. **195**, 851–860 (2018)

13. E. Bernal, D. Edgar, B. Burnes, Building sustainability on deep values through mindfulness nurturing. Ecol. Econ. **146**, 645–657 (2018)

14. L. Guardigli, M. Bragadin, F. Della Fornace, C. Mazzoli, D. Prati, Energy retrofit alternatives and cost-optimal analysis for large public housing stocks. Energy Build. **166**, 48–59 (2018)

15. B. Huang, X. Wang, H. Kua, Y. Geng, R. Bleischwitz, J. Ren, Construction and demolition waste management in China through the 3R principle. Resour. Conserv. Recycl.. Conserv. Recycl. **129**, 36–44 (2018)

16. S.A. Miller, V. John, S. Pacca, A. Horvath, Carbon dioxide reduction potential in the global cement industry by 2050. Cem. Concr. Res.. Concr. Res. **114**, 115–124 (2018)

17. J. Fořt, R. Černý, Transition to circular economy in the construction industry: environmental aspects of waste brick recycling scenarios. Waste Manage. **118**, 510–520 (2020)

18. R. Agrawal, V. Wankhede, A. Kumar, A. Upadhyay, J. Garza-Reyes, Nexus of circular economy and sustainable business performance in the era of digitalization. Int. J. Product. Perform. Manag.Manag. **71**(3), 748–774 (2021)

19. M. Ghufran, K. Khan, F. Ullah, A. Nasir, A. Al Alahmadi, A. Alzaed, M. Alwetaishi, Circular economy in the construction industry: a step towards sustainable development. Buildings **12**(7), 1004 (2022)

20. UN, in *Report of the World Commission on Environment and Development: Our Common Future,* 1987. [Online]. Available: http://www.un-documents.net/our-common-future.pdf. Accessed 08 Sept 2023

21. M. Zu Castell-Rüdenhausen, M. Wahlström, T. Fruergaard Astrup, C. Jensen, A. Oberender, P. Johansson, E. Waerner, Policies as drivers for circular economy in the construction sector in the Nordics. Sustainability **13**(16), 9350 (2021)

22. G. Benachio, M. Freitas, S. Tavares, Circular economy in the construction industry: a systematic literature review. J. Clean. Prod. **260**, 121046 (2020)

23. Y. Kalmykova, M. Sadagopan, L. Rosado, Circular economy—from review of theories and practices to development of implementation tools. Resour. Conserv. Recycl.. Conserv. Recycl. **135**, 190–201 (2018)

24. F. Ullah, S. Sepasgozar, C. Wang, A systematic review of smart real estate technology: drivers of, and barriers to, the use of digital disruptive technologies and online platforms. Sustainability **10**(9), 3142 (2018)

25. C. Garcés-Ayerbe, P. Rivera-Torres, I. Suárez-Perales, D. Leyva-De La Hiz, Is it possible to change from a linear to a circular economy? An overview of opportunities and barriers for European small and medium-sized enterprise companies. Int. J. Environ. Res. Public Health **16**(5), 851 (2019)

26. T. Joensuu, H. Edelman, A. Saari, Circular economy practices in the built environment. J. Clean. Prod. **276**, 124215 (2020)

27. P. Ghisellini, C. Cialani, S. Ulgiati, A review on circular economy: the expected transition to a balanced interplay of environmental and economic systems. J. Clean. Prod. **114**, 11–32 (2016)
28. D. Masi, S. Day, J. Godsell, Supply chain configurations in the circular economy: a systematic literature review. Sustainability **9**(9), 1602 (2017)
29. K. Hobson, in *Trajectories in Environmental Politics*, ed. by G. Hayes, S. Jinnah, P. Kashwan, D.M. Konisky, S. Macgregor, J.M. Meyer, A.R. Zito (Routledge, London, 2022), p. 324
30. T. Hák, S. Janoušková, B. Moldan, Sustainable development goals: a need for relevant indicators. Ecol. Ind. **60**, 565–573 (2016)
31. A. Wieser, M. Scherz, S. Maier, A. Passer, H. Kreiner, in *Implementation of Sustainable Development Goals in Construction Industry—A Systemic Consideration of Synergies and Trade-offs*. IOP Conference Series: Earth and Environmental Science, (Graz, Austria, 2019)
32. B. Xia, A. Olanipekun, Q. Chen, L. Xie, Y. Liu, Conceptualising the state of the art of corporate social responsibility (CSR) in the construction industry and its nexus to sustainable development. J. Clean. Prod. **195**, 340–353 (2018)
33. P. Mansell, S. Philbin, E. Konstantinou, Delivering UN sustainable development goals' impact on infrastructure projects: an empirical study of senior executives in the UK construction sector. Sustainability **12**(19), 7998 (2020)
34. P. Howden-Chapman, J. Siri, E. Chisholm, R. Chapman, C. Doll, A. Capon, SDG 3: Ensure healthy lives and promote wellbeing for all at all ages, 2017. [Online]. Available: https://council.science/wp-content/uploads/2017/03/SDGs-interactions-3-healthy-lives.pdf. Accessed 08 Sept 2023
35. K. Chaudhary, P. Vrat, Circular economy model of gold recovery from cell phones using system dynamics approach: a case study of India. Environ. Dev. Sustain. **22**, 173–200 (2020)
36. S. Türkeli, R. Kemp, B. Huang, R. Bleischwitz, W. McDowall, Circular economy scientific knowledge in the European union and China: a bibliometric, network and survey analysis (2006–2016). J. Clean. Prod. **197**(1), 1244–1261 (2018)
37. A. Wijewansha, G. Tennakoon, K. Waidyasekara, B. Ekanayake, Implementation of circular economy principles during pre-construction stage: the case of Sri Lanka. Built Environ. Proj. Asset Manage. **11**(4), 750–766 (2021)
38. M. Ormazabal, V. Prieto-Sandoval, R. Puga-Leal, C. Jaca, Circular economy in Spanish SMEs: challenges and opportunities. J. Clean. Prod. **185**, 157–167 (2018)
39. O. Hanseth, K. Lyytinen, Design theory for dynamic complexity in information infrastructures: the case of building internet. J. Inf. Technol. **25**(1), 1–19 (2010)
40. S. Trivedi, K. Snehal, B. Das, S. Barbhuiya, A comprehensive review towards sustainable approaches on the processing and treatment of construction and demolition waste. Constr. Build. Mater. **393**, 132125 (2023)
41. Z. Wu, A.T. Yu, L. Shen, G. Liu, Quantifying construction and demolition waste: an analytical review. Waste Manage. **34**(9), 1683–1692 (2014)
42. Official Journal of the European Union, in *2014/955/EU: Commission Decision of 18 December 2014 Amending Decision 2000/532/EC on the List of Waste Pursuant to Directive 2008/98/EC of the European Parliament and of the Council Text with EEA Relevance*, 30 Dec 2014. [Online]. Available: https://eur-lex.europa.eu/legal-content/EN/TXT/?uri=celex%3A32014D0955. Accessed 10 Sept 2023
43. Journal of Laws of 2021, in *The Act of 17 November 2021 Amending the Act on Waste and Some Other Acts* (Item 2151). [Online]. Available: https://isap.sejm.gov.pl/isap.nsf/DocDetails.xsp?id=WDU20210002151. Accessed 10 Sept 2023
44. Regulation of the Minister of Environment, in *Natural Resources and Forestry of 24 December 1997 on the Classification of Waste* [Online]. Available: https://www.prawo.pl/akty/dz-u-1997-162-1135,16799265.html. Accessed 10 Sept 2023
45. Official Journal EUC, in *Commission Notice on Technical Guidelines on the Classification of Waste* (Luxembourg, 2018) (2018/C 124/01). [Online]. Accessed 10 Sept 2023
46. M. Norouzi, M. Chàfer, L.F. Cabeza, L. Jiménez, D. Boer, Circular economy in the building and construction sector: a scientific evolution analysis. J. Build. Eng. **44**, 102704 (2021)

47. M. Bravo, J. Brito, J. Pontes, L. Evangelista, Mechanical performance of concrete made with aggregates from construction and demolition waste recycling plants. J. Clean. Prod. **99**, 59–74 (2015)
48. United Nations Environment Programme, in *Annual report 2021*, 2022. [Online]. Available: https://wedocs.unep.org/bitstream/handle/20.500.11822/37946/UNEP_AR2021.pdf. Accessed 05 Sept 2023
49. J. Rodrigue, C. Comtois, B. Slack, *The Geography of Transport Systems* (Routledge, London, 2016)
50. J. Liu, H. Mooney, V. Hull, S. Davis, J. Gaskell, T. Hertel, J. Lubchenco, K. Seto, P. Gleick, C. Kremen, S. Li, Systems integration for global sustainability. Science **347**, 6225 (2015)
51. L. Wang, X. Xue, Z. Zhao, Z. Wang, The impacts of transportation infrastructure on sustainable development: emerging trends and challenges. Int. J. Environ. Res. Public Health **15**(6), 1172 (2018)
52. D. Kammen, D.A. Sunter, City-integrated renewable energy for urban sustainability. Science **352**(6288), 922–928 (2016)
53. L. Kraus, H. Proff, Sustainable urban transportation criteria and measurement—a systematic literature review. Sustainability **13**(13), 7113 (2021)
54. United Nations, in *Sustainable Transport, Sustainable Development: Interagency Report for Second Global Sustainable Transport Conference* (2021). [Online]. Available: https://sdgs.un. org/sites/default/files/2021-10/Transportation%20Report%202021_FullReport_Digital.pdf. Accessed 05 Sept 2023
55. European Environment Agency, in *Decarbonising Road Transport—The Role of Vehicles, Fuels and Transport Demand* (Transport and Environment report 2021). [Online]. Available: https://www.eea.europa.eu/publications/transport-and-environment-report-2021. Accessed 05 Sept 2023
56. European Environment Agency, in *Digitalisation in the Mobility System: Challenges and Opportunities* (Transport and Environment report 2022). [Online]. Available: https://www.eea.europa.eu/publications/transport-and-environment-report-2022/transport-and-environment-report/view. Accessed 05 Sept 2023
57. C. Jeon, A. Amekudzi-Kennedy, R. Guensler, Evaluating plan alternatives for transportation system sustainability: Atlanta metropolitan region. Int. J. Sustain. Transp. **4**(4), 227–247 (2010)
58. A. Brinch, S. Hansen, N. Hartmann, A. Baun, EU regulation of nanobiocides: challenges in implementing the biocidal product regulation (BPR). Nanomaterials **6**(2), 33 (2016)
59. G. Ševčenko-Kozlovska, K. Čižiūnienė, The impact of economic sustainability in the transport sector on GDP of neighbouring countries: following the example of the Baltic states. Sustainability **14**(6), 3326 (2022)
60. I. Khan, K. Saeed, Nanoparticles: properties, applications and toxicities. Arab. J. Chem. **12**(7), 908–931 (2019)
61. M. Gómez Garzón, Nanomateriales, nanopartículas y síntesis verde. Revista Repertorio de Medicina y Cirugía. Fundación Universitaria de Ciencias de la Salud—FUCS (2018)
62. W. Shin, J. Cho, A.G. Kannan, Y. Lee, D. Kim, Cross-linked composite gel polymer electrolyte using mesoporous methacrylate-functionalized SiO_2 nanoparticles for lithium-Ion polymer batteries. Sci. Rep. **6**, 26332 (2016)
63. V. Makarov, A. Love, O. Sinitsyna, S. Makarova, I. Yaminsky, M. Taliansky, N. Kalinina, "Green" nanotechnologies: Synthesis of metal nanoparticles using plants. **6**(1), 35–44 (2014)
64. A. Boroumand Moghaddam, F. Namvar, M. Moniri, P. M. Tahir, S. Azizi, R. Mohamad, Nanoparticles biosynthesized by fungi and yeast: a review of their preparation, properties, and medical applications. Molecules **20**(9), 16540–16565 (2015)

Chapter 2
Life Cycle Assessment (LCA) Fundamental Principles

Esperanza Conradi-Galnares, Begoña Blandón-González, and Madelyn Marrero-Meléndez

Abstract This second chapter systematically acquaints the reader with Life Cycle Assessment (LCA), meticulously defining its primary objectives, scope, and application in the construction sector. This chapter also delineates the normative framework governing Life Cycle Assessment, accompanied by an exploration of its specific terminology.

In our society, buildings are ubiquitous, yet they inherently entail adverse environmental impacts. As previously discussed, throughout their lifecycle, they deplete significant resources and energy, encroach upon land, and eventually face demolition. With a burgeoning interest in environmental concerns, particularly within the construction sector, there is a heightened focus on sustainable housing technologies and construction practices [1].

Frequently, products marketed as inexpensive in the short term may incur substantial maintenance or waste management expenses, while highly technological products can have exorbitant production costs that are never recovered. Adopting a life cycle perspective is crucial for informed decision-making when selecting the most suitable technology and minimizing the environmental footprint of buildings through design or refurbishment [2]. While the emphasis is often placed on

Present Address:
E. Conradi-Galnares (✉) · B. Blandón-González
Department of Building Construction I, Higher Technical School of Architecture, University of Seville, Seville, Spain
e-mail: conradi@us.es

B. Blandón-González
e-mail: bblandon@us.es

Present Address:
M. Marrero-Meléndez
Department of Building Construction II, Higher Technical School of Building Engineering, University of Seville, Seville, Spain
e-mail: madelyn@us.es

© The Author(s) 2025
P. Mercader-Moyano and P. Porras-Pereira (eds.), *Life Cycle Analysis Based on Nanoparticles Applied to the Construction Industry*,
https://doi.org/10.1007/978-3-031-79115-4_2

reducing energy consumption and utilizing eco-friendly materials, the significance of life-cycle thinking is increasingly recognized [1].

1 Life Cycle Assessment: Definition and Objectives

Sustainable construction processes have been of key importance to the construction industry for the last two decades. The environmental impacts that may result from such processes are of equal importance to the decision makers. This has led to the quest for the application of life cycle thinking, with a view to reuse and recycles materials [3].

Building materials play a crucial role in the construction industry, and the use of sustainable building materials has become increasingly important in recent years. Life cycle assessment (LCA) is a widely used method to assess environmental burdens of a product, service or process throughout their entire life cycle, from the extraction of raw materials to their disposal, from cradle to grave, in order to assist consumers in making decisions that will benefit the environment (Fig. 1).

The main goals of LCA are to quantify or otherwise characterize all the inputs and outputs over a product's life cycle, specify the potential environmental impacts of these material flows and consider alternative approaches that change those impacts for the better (Fig. 2) [5].

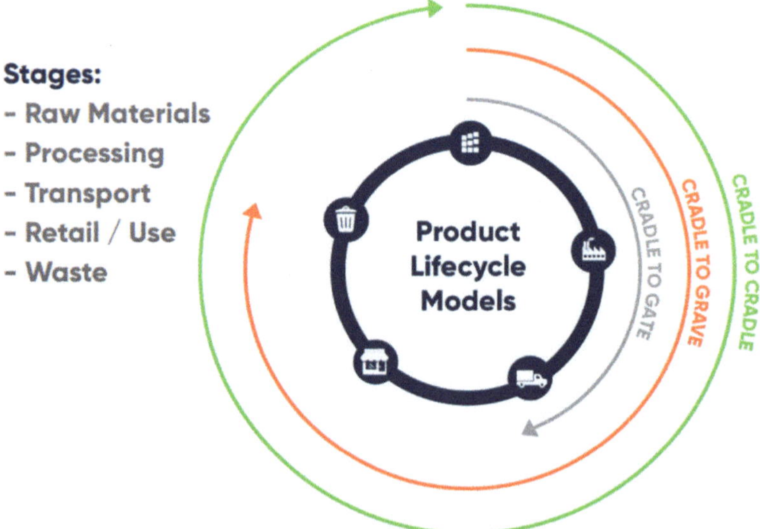

Fig. 1 Product's life cycle [4]

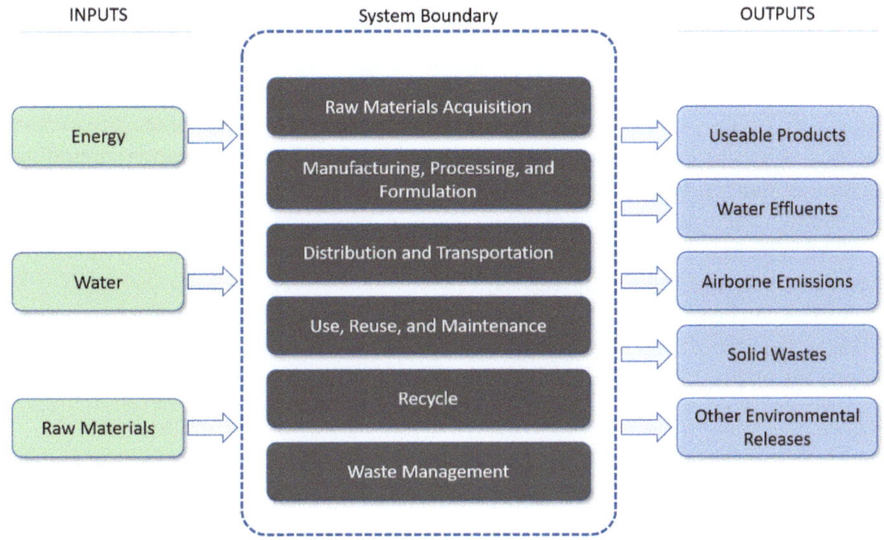

Fig. 2 Inputs and outputs over a product's life cycle [5]

Incorporating considerations of the natural environment, human health, and resource depletion, along with the life cycle perspective, LCA ensures that problem-shifting between various life cycle stages, regions, and environmental issues is avoided [1]. LCA studies help to avoid resolving one environmental problem while creating others. Life Cycle Assessment is therefore a powerful decision support tool necessary to make consumption and production more sustainable [4]. LCA can help identify the areas of the life cycle where environmental impacts can be reduced and promote sustainable production practices [1], although in each life cycle stage there is the potential to reduce resource consumption and improve the performance of products [4].

Life-cycle assessment has emerged as a valuable tool for decision support, benefiting both policymakers and industries in evaluating the cradle-to-grave impacts of products or processes. This evolution is driven by three main factors. Firstly, government regulations are shifting towards "life-cycle accountability," emphasizing that manufacturers are responsible not only for direct production impacts but also for those associated with product inputs, use, transportation, and disposal. Secondly, businesses are engaging in voluntary initiatives incorporating LCA and product stewardship components, such as ISO 14000 and the Chemical Manufacturer Association's Responsible Care Program, to promote continuous improvement through enhanced environmental management systems. Thirdly, environmental "preferability" has become a criterion in consumer markets and government procurement guidelines, further emphasizing the importance of considering environmental impacts throughout the product lifecycle Together these developments have placed LCA in a central role as a tool for identifying cradle-to-grave impacts both of products and the materials from which they are made [6].

The impacts of all life cycle stages need to be considered comprehensively by citizens, companies, and governments, when making decisions regarding consumption and production patterns, policies, and management strategies. On a macro level, life cycle approaches prevent the shifting of problems from one stage of the life cycle to another, from one geographic area to another, and from one environmental medium, such as air quality, to another, such as water or land. On a micro level, they empower individuals, such as product designers, service providers, and government agents, to make informed choices for the long term, taking into account all environmental media, such as air, water, and land [4].

A LCA can be performed to a building even before the start of construction activity to know the impacts on the environment, thereby giving the contractor a scope to reduce the environmental impacts by using alternate construction materials [7].

2   LCA in Construction Sector

In industrial processes, LCA is widely utilized and frequently employed to assess the environmental impact of products and processes [5, 7]. Buildings, however, are special products that differ thoroughly from these mostly controlled processes. In the construction industry, such a study is therefore on the average much more complex because of multiple issues [1]:

- Buildings and constructions have extremely long lifetimes, often more than 50 years. It is difficult to predict the life cycle 'from cradle-to-grave', there are more uncertainty variables and parameters [8].
- During this life span, the building or construction may undergo many changes in its form and function because of maintenance and retrofitting or because the shorter lifespan of some of its elements and components. These changes can be as significant, or even more significant, than the original construction. The ease with which changes can be made and the opportunity to minimise the environmental effects of changes are partly functions of the original design [8].
- Many of the environmental impacts of a building occur during its use (energy and water consumption). Proper design and material selection are critical to minimise those in-use environmental loads, being crucial to take into consideration the distances to factories [8].
- There are many actors in the building and construction column. The designer, who makes the decisions about the final building or construction or its required performance, does not produce the components, nor does he or she build the building neither the construction. In many cases, very formal relations exist among commissioners, designers, and contractors [8].
- Traditionally, each building or construction is unique and is designed as such. There is very little standardisation in the building design and, usually, many different materials and processes are used in it. Innovative choices have to be made for each specific situation [8].

The long lifespan and dependence of user behaviour thus require much more assumptions, coming with larger uncertainties and consequently influence the credibility of the results [9]. So, since the building process is less standardized than industrial processes, such a Life Cycle Assessment is a challenging task [1].

LCA studies of entire buildings are a great tool to investigate building concepts and to support decision-making to reduce environmental burdens. Nonetheless, the LCA methodology does has inherent limitations, thus the results should be interpreted and applied cautiously. Firstly, comparing cases is challenging due to their specific characteristics such as layout, climate, comfort requirements, local regulations, and so forth. Additionally, the widespread estimations of building lifespans represent another limitation. While these limitations can be partially mitigated by calculating annual burdens per square meter of useful floor space or per person, other aspects of the studies may vary, including system boundaries, assumptions, level of detail, LCIA methods, and so on. Moreover, LCA is essentially a model and simplification of reality, necessitating the making of assumptions that can introduce uncertainties at various levels: model, scenario, and parameter uncertainties. Processing the first two aspects statistically is often challenging and they are frequently excluded from analysis. However, with the latter, this is feasible as data quality indicators are accessible for all materials and processes in the databases. Parameter uncertainty is also frequently compounded by data gaps, leading to less precise data being utilized. When accounting for the variability and stochastic error of the figures, reliability is bolstered, but interpretation necessitates the use of probability statements, which are less common. Nonetheless, useful conclusions can still be drawn [1].

As previously stated, the utilization phase of buildings overwhelmingly contributes to the environmental burdens throughout their entire life cycle, primarily due to high energy consumption. The burdens of this phase are based on estimations, taking average values of the whole society into account. Since individual inhabitant behaviour is difficult to predict, it is also an issue of concern when considering the reliability of any conclusion on energy consumption. This reality underscores the practical significance of LCA, regardless of the precision of the calculations. Research concluded that numerous efficiency enhancements fail to decrease energy consumption as significantly as anticipated. As they make energy services cheaper, the demand for these services will increase. For example, if a dwelling is well insulated, residents are more likely to heat up the spaces above the calculated temperature, since this entails only a limited additional cost. This psychological phenomenon is called the rebound effect and until now this has not been taken into account [1, 10]. A stochastic approach based on real data could partly counter this problem; however, rebound effects will always occur as economic savings will trigger other (non-building related) expenditures which of course entail also environmental burdens. An extra difficulty is the fact that user behaviour and consumption habits are often regionally defined [1, 11].

Another limitation of current LCA practice within the construction sector is its isolated approach to environmental issues. Often, the focus is restricted to seeking environmental optimizations without linking them to other factors. For instance, LCA typically does not consider quality, energy, structural, or aesthetic requirements.

Furthermore, while design's impact on the environmental profile is significant, it is often overlooked compared to purely technological improvements. Additionally, financial feasibility is seldom taken into consideration, despite the availability of tools like Life Cycle Costing. Although new regulations and frameworks have been developed for assessing all aspects of sustainability, they are not yet widely implemented [1].

Indeed, despite certain limitations, the LCA technique remains a powerful and science-based tool for evaluating environmental impacts.

3 Life Cycle Assessment Terminology

The following list of terms includes the main terms needed to understand when reading, evaluating, or interpreting an LCA report. For the purposes of this document, the following terms and definitions, included in the ISO 14044:2006 standard, apply [12–14].

- *Life cycle*

Consecutive and interlinked stages of a product system, from raw material acquisition or generation from natural resources to final disposal [12].

- *Life Cycle Assessment (LCA)*

Compilation and evaluation of the inputs, outputs, and the potential environmental impacts of a product system throughout its life cycle [12, 14].

- *Life cycle inventory analysis (LCI)*

Phase of life cycle assessment involving the compilation and quantification of inputs and outputs for a product throughout its life cycle [12].

- *Life cycle impact assessment (LCIA)*

Phase of life cycle assessment aimed at understanding and evaluating the magnitude and significance of the potential environmental impacts for a product system throughout the life cycle of the product [12, 14].

- *Life cycle interpretation*

Phase of life cycle assessment in which the findings of either the inventory analysis or the impact assessment, or both, are evaluated in relation to the defined goal and scope in order to reach conclusions and recommendations [12, 14].

- *Comparative assertion*

Environmental claim regarding the superiority or equivalence of one product versus a competing product that performs the same function [12].

- *Environmental aspect*

Element of an organization's activities, products or services that can interact with the environment [12].

- *Co-product*

Any of two or more products coming from the same unit process or product system [12].

- *Process*

Set of interrelated or interacting activities that transforms inputs into outputs [12].

- *Elementary flow*

Material or energy entering the system being studied that has been drawn from the environment without previous human transformation, or material or energy leaving the system being studied that is released into the environment without subsequent human transformation [12].

- *Energy flow*

Input to or output from a unit process or product system, quantified in energy units [12].

- *Embodied energy*

Total energy expended throughout the entire lifecycle of a product, encompassing all stages from raw material extraction and processing to manufacturing, transportation, use, and disposal. It reflects the cumulative energy required to bring a product into its current state [13].

- *Operating energy*

Total energy required to operate a system, such as the total energy consumed by a building during its lifetime [13].

- *Raw material*

Primary or secondary material that is used to produce a product [12].

- *Ancillary input*

Material input that is used by the unit process producing the product, but does not constitute part of the product [12].

- *Functional unit*

Quantified performance of a product system for use as a reference unit [12]. It provides a quantitative expression of the product under study, encompassing its function, quantity, performance, and duration. The functional unit remains consistent across the different alternatives being evaluated in the study [13].

- *Input*

  Product, material, or energy flow that enters a unit process [12].

- *Intermediate flow*

  Product, material, or energy flow occurring between unit processes of the product system being studied [12].

- *Intermediate product*

  Output from a unit process that is input to other unit processes that require further transformation within the system [12].

- *Life cycle inventory analysis result (LCI)*

  Result outcome of a life cycle inventory analysis that catalogues the flows crossing the system boundary and provides the starting point for life cycle impact assessment [12].

- *Output*

  Product, material or energy flow that leaves a unit process [12].

- *Product system*

  Collection of unit processes with elementary and product flows, performing one or more defined functions, and which models the life cycle of a product [12].

- *Sensitivity analysis*

  Systematic procedures for estimating the effects of the choices made regarding methods and data on the outcome of a study [12].

- *System boundary*

  Set of criteria specifying which unit processes are part of a product system [12, 14]. They serve as an interface between a product system and the environment or other product systems. It defines the activities and processes that will be included in each life cycle stage for the LCA analysis and those that will be excluded [13].

- *Uncertainty analysis*

  Systematic procedure to quantify the uncertainty introduced in the results of a life cycle inventory analysis due to the cumulative effects of model imprecision, input uncertainty and data variability [12].

- *Unit process*

  Smallest element considered in the life cycle inventory analysis for which input and output data are quantified [12].

- *Category endpoint*

  Attribute or aspect of natural environment, human health, or resources, identifying an environmental issue giving cause for concern [12].

- *Characterization factor*

Factor derived from a characterization model which is applied to convert an assigned life cycle inventory analysis result to the common unit of the category indicator [12].

- *Environmental mechanism*

System of physical, chemical, and biological processes for a given impact category, linking the life cycle inventory analysis results to category indicators and to category endpoints [12].

- *Impact category*

Class representing environmental issues of concern to which life cycle inventory analysis results may be assigned [12]. These categories encompass specific areas related to ecosystem quality, resource use, or human health that are quantified in the Life Cycle Assessment [12].

- *Impact category indicator*

Quantifiable representation of an impact category [12]

- *Completeness check*

Process of verifying whether information from the phases of a life cycle assessment is sufficient for reaching conclusions in accordance with the goal and scope definition [12].

- *Consistency check*

Process of verifying that the assumptions, methods and data are consistently applied throughout the study and are in accordance with the goal and scope definition performed before conclusions are reached [12].

- *Sensitivity check*

Process of verifying that the information obtained from a sensitivity analysis is relevant for reaching the conclusions and giving recommendations [12].

- *Evaluation*

Element within the life cycle interpretation phase intended to establish confidence in the results of the life cycle assessment [12].

- *Critical review*

Process intended to ensure consistency between a life cycle assessment and the principles and requirements of the international standards on life cycle assessment [12].

- *Interested party*

Individual or group concerned with or affected by the environmental performance of a product system, or by the results of the life cycle assessment [12].

4   Normative Frame of Reference for LCA

In recent years, many countries have introduced normative criteria for the LCA of building materials to regulate their environmental impact. The development of normative criteria can vary depending on factors such as national policies, regulations, and industry practices. Therefore, it is important to conduct comparative studies to understand the differences and similarities in the normative criteria for LCA of building materials in different countries [1].

The normative frame of Life Cycle Assessment is a vital tool for making Informed decisions that contribute to sustainable, environmentally responsible design and construction. By following the systematic process of LCA and considering the normative implications of their choices, experts can play a crucial role in minimizing the environmental footprint of the built environment.

Although ISO standards outline the global framework for LCA, the specific technique for calculating environmental impacts is not prescribed. Depending on the nature of the research, various methods can be selected, each defined by its environmental mechanisms as described in ISO 14044:2006 related to environmental management, Life Cycle Assessment, and requirements and guidelines [1].

Following, a list of existing regulations regarding Life Cycle Assessment (LCA) of building materials is presented [15].

- ISO 14040 and ISO 14044 on life cycle assessment

These are the prominent international standards governing Life Cycle Assessment (LCA). They primarily centre on the process of conducting LCA, tracking a product's impact from cradle to grave. ISO 14040 establishes the principles and framework for conducting Life Cycle Assessment, while ISO 14044 specifies the requirements and provides guidelines for implementing LCA [15].

- ISO 14067 on the carbon footprint of products

ISO 14067 outlines principles, requirements, and guidelines for quantifying and reporting the carbon footprint of a product, specifically its impact on climate change. However, carbon offsetting and communication of carbon footprint are not addressed within the scope of this standard [15].

The standard is consistent with ISO 14040 and ISO 14044. However, it also includes requirements on specific issues relevant to carbon footprinting, including land-use change, carbon uptake, biogenic carbon emissions, and soil carbon change [15].

- ISO 14020, ISO 14021, ISO 14024, ISO 14025, and ISO 14026 on environmental labels

These standards set out principles, requirements and guidelines for the development and use of environmental labels and declarations, as well as for the communication of footprint information [15].

In addition to the generic product standards and guidelines aforementioned, there are more-product-specific international standards [15].

- ISO 22526 parts 1, 2, 3 and 4 on carbon footprint and removals for biobased plastics

The ISO 22526 series, parts 1, 2, and 3, sets out the principles, requirements, and guidelines for the quantification of, and reporting on, the carbon footprint throughout the life cycle of partly or wholly biobased plastics. The process carbon footprinting of biobased plastics is carried out in accordance with the ISO 14067 guidelines on carbon footprinting. The series also establishes the standard for determining the material carbon footprint, including the carbon dioxide removed from the air and incorporated into a given plastic product [15].

- ISO 20915 on life cycle inventory for steel products

ISO 20915 delineates requirements and guidelines for conducting life cycle inventory (LCI) studies specifically tailored to steel products, encompassing closed-loop recycling. The standard adheres to the principles of international LCA standards ISO 14040 and ISO 14044, illustrating how these principles can be effectively applied to the manufacturing and recycling processes of steel products [15].

Moreover, there are also Standards and guidelines available for corporations and other organizations, such as:

- ISO 14063 on environmental communication

ISO 14063 provides guidelines to organizations regarding general principles, policy, strategy, and activities pertaining to both internal and external environmental communication. It is applicable to all organizations, irrespective of their size, type, location, structure, activities, products, and services. ISO 14063 can be utilized in conjunction with any of the ISO 14000 family of standards, or independently [15].

- ISO 14064, parts 1, 2 and 3, on greenhouse gases and removals

The ISO 14064 series outlines the principles and requirements for quantifying, monitoring, reporting, verifying, and validating activities intended to reduce greenhouse gas emissions or enhance removal, at both the organization and project levels. These standards specify requirements for greenhouse gas (GHG) project planning and offer guidelines for identifying GHG sources, sinks, and reservoirs. Part 1 of the series addresses the organization level, while Part 2 focuses on the project level. Part 3 provides guidance for verifying and validating GHG statements [15].

- ISO 14080 on greenhouse gas management and climate actions

ISO 14080 establishes guidelines and principles for identifying, assessing, revising, developing, and managing methodologies for actions addressing climate change, encompassing both adaptation to its impacts and mitigation of greenhouse gas emissions [15].

- ISO 14069 on guidance for the application of ISO 14064-1

ISO 14069 delineates the principles, concepts, and methods pertaining to the quantification and reporting of both direct and indirect greenhouse gas emissions for an organization. It offers guidance for quantifying and reporting direct emissions, energy indirect emissions, and other indirect emissions [15].

- ISO 14072 on effective application of ISO 14040/44 to organizations

ISO 14072 is a technical specification that sets out additional requirements and guidelines for an effective application of ISO 14040 and ISO 14044 to organizations. Among other things, it details the application of LCA principles and methodology to organizations, the benefits that LCA can bring to organizations, by applying the LCA methodology at organizational level, the system boundary, specific considerations when dealing with LCI, LCIA, and interpretation, and the limitations regarding reporting, environmental declarations, and comparative assertions [15].

Finally, there are combined standards and guidelines that contain information for environmental assessments at the product, organization, and country levels [15].

- ISO 14046 on water footprint

ISO 14046 establishes principles, requirements, and guidelines concerning the assessment of water footprints for products, processes, and organizations using LCA methodology. This standard aligns with the international ISO 14044 LCA standard [15].

- Global Water Footprint Standard

The global water footprint standard in the Water Footprint Assessment Manual offers guidelines on the water footprint of individual processes and products, on carrying out water-footprint sustainability assessments, and on using the results to prioritize strategic actions to be taken at the local, national, regional, and global levels [15].

- ISO 14001, 14006, 14007, 14008, and 14009 on environmental management systems

The ISO 14000 series, which includes standards like 14001, 14006, 14007, 14008, and 14009, establishes requirements and guidelines for environmental management systems within organizations. Aligned with the organization's environmental policy, the primary objectives of an environmental management system include enhancing environmental performance, meeting compliance obligations, and achieving environmental objectives [15].

Additionally, the standards provide guidelines and principles for eco-design, quantifying environmental costs and benefits, evaluating environmental impacts in monetary terms, and integrating material circulation into the design and development of products [15].

- Greenhouse gas protocol standards at the product and organization levels

The Greenhouse Gas Protocol offers standards for quantifying life cycle greenhouse gas emissions at both the product and organization levels. The Product Life Cycle Accounting and Reporting Standard offers guidelines for organizations to assess the life cycle greenhouse gas emissions of their products. The Corporate Accounting and Reporting Standard outlines the requirements for companies to document their greenhouse gas emissions. The Corporate Value Chain Accounting and Reporting Standard provides guidance to companies on measuring emissions throughout their entire value chains, encompassing emissions beyond their operations and electricity consumption [15].

Nowadays, there are also sustainability standards that currently are in the development phase, although some of them are on the publication stage, such as:

- ISO 14068 on carbon neutrality

ISO 14068 will serve as a standard for organizations, constituting a part of the previously mentioned 14060 family of standards. Its primary emphasis will be on GHG management and achieving carbon neutrality [15].

- ISO 59004, 59010, 59020 on the circular economy

This ISO series (ISO 59004, 59010, 59020) will set out guidelines and principles for implementing and measuring circularity and for circular business models and value chains [15].

- ISO 14074 on life cycle assessment on normalization, weighting, and interpretation

ISO 14074 will establish guidelines, requirements, and principles for the normalization, weighting, and interpretation of LCAs [15].

- ISO 14097 on greenhouse gas reporting for financing activities

ISO 14097 will set out principles and requirements for assessing and reporting on financing activities and investments related to climate change [15].

Normative criteria for Life Cycle Assessment play an essential role in providing a structured and consistent framework for appraising the environmental impacts of products and processes, encompassing building materials. However, it is imperative to acknowledge that these criteria carry their own set of limitations and implications, which necessitate consideration during the execution of LCAs.

Given the substantial reliance of Life Cycle Assessment on data, the challenges associated with securing accurate and representative data, particularly for niche or innovative materials, can pose a significant hurdle. Inaccurate or incomplete data introduces a risk of unreliable LCA results, potentially influencing the decision-making process. Furthermore, the diverse choices in defining system boundaries can yield markedly disparate outcomes. Therefore, a meticulous definition of system boundaries in LCA is imperative to ensure a faithful representation of the environmental impacts associated with building materials.

Additionally, the allocation of environmental impacts among multiple products or processes introduces subjectivity. Awareness of the chosen allocation methods and their implications is crucial when comparing materials or processes. Aligning the impact categories considered with project-specific goals and priorities becomes pivotal in this context.

A critical limitation lies in regional variability, as LCA criteria may inadequately address regional disparities in material sourcing, energy utilization, or waste management. Recognizing the regional context becomes indispensable when interpreting LCA results, as it allows for the incorporation of variations in environmental impacts associated with specific regions.

In summary, the interpretation of Life Cycle Assessment results must be grounded in the particular context of project goals and environmental priorities to facilitate informed and sustainable decision-making.

References.

1. M. Buyle, J. Braet, A. Audenaert, Life cycle assessment in the construction sector: a review. Renew. Sustain. Energy Rev. **26**, 379–388 (2013)
2. I. Zabalza Bribián, A. Valero Capilla, A. Aranda Usón, Life cycle assessment of building materials: comparative analysis of energy and environmental impacts and evaluation of the eco-efficiency improvement potential. Build. Environ. **46**(5), 1133–1140 (2011)
3. K. Yahya, H. Boussabaine, A. Alzaed, Using life cycle assessment for estimating environmental impacts and eco-costs from the metal waste in the construction industry. Manage. Environ. Qual. **27**(2), 227–244 (2016)
4. Ecochain, in *Life Cycle Assessment (LCA)—Complete Beginner's Guide* [Online]. Available. https://ecochain.com/blog/life-cycle-assessment-lca-guide/. Accessed 06 Sept 2023
5. Sustainable Facilities Tool, in *Life Cycle Assessment—GSA Sustainable Facilities Tool* [Online]. Available. https://sftool.gov/plan/400/life-cycle-assessment. Accessed 05 Sept 2023
6. United Nations Environment Programme, in *What is Life Cycle Thinking?* [Online]. Available. https://www.lifecycleinitiative.org/activities/what-is-life-cycle-thinking/. Accessed 06 Sept 2023
7. Y. Pamu, V. Kumar, M. Shakir, H. Ubbana, Life cycle assessment of a building using open-LCA software. Mater. Today: Proc **52**(3), 1968–1978 (2022)
8. S. Kotaji, A. Schuurmans and S. Edwards, in *Life-Cycle Assessment in Building and Construction: A state-of-the-Art Report*, ed. by S. Kotaji, A. Schuurmans, S. Edwards, Society of Environmental Toxicology and Chemistry SETAC, 2003
9. I. Blom, L. Itard, A. Meijer, Environmental impact of building-related and user-related energy consumption in dwellings. Build. Environ. **46**(8), 1657–1669 (2011)
10. L.A. Greening, D.L. Greene, C. Difiglio, Energy efficiency and consumption—the rebound effect—a survey. Energy Policy **28**(6–7), 389–401 (2000)
11. O. Ortiz-Rodríguez, F. Castells, G. Sonnemann, Life cycle assessment of two dwellings: one in Spain, a developed country, and one in Colombia, a country under development. Sci. Total. Environ. **408**(12), 2435–2443 (2010)
12. International Organization for Standardization (ISO): ISO 14044:2006, in *Environmental Management—Life Cycle Assessment—Requirements and Guidelines* [Online]. Available. https://www.iso.org/obp/ui/#iso:std:iso:14044:ed-1:v1:en. Accessed 09 Sept 2023

13. Sustainable Facilities Tool, in *LCA Key Terms* [Online]. Available. https://sftool.gov/plan/630/lca-key-terms. Accessed 09 Sept 2023
14. United Nations Environment Programme, in *Glossary of Life Cycle Terms* [Online]. Available. https://www.lifecycleinitiative.org/activities/life-cycle-terminology-2/. Accessed 09 Sept 2023
15. G. Pallas, in *Life Cycle-Based Sustainability Standards and Guidelines*, 2021 (PRé Sustainability). [Online]. Available. https://pre-sustainability.com/articles/lca-standards-and-guidelines/. Accessed 09 Sept 2023

Chapter 3
Life Cycle Assessment (LCA) Methodology

Marco Antonio Sánchez-Burgos, Esperanza Conradi-Galnares, and Dimitrios Dragatogiannis

Abstract This third chapter meticulously articulates the Life Cycle Assessment methodology, providing a detailed exposition of the various phases inherent in the process, an exploration of diverse impact categories, and an overview of commonly utilized tools for LCA development. Additionally, the chapter sheds light on prevalent environmental certification standards. Furthermore, the chapter concludes by presenting insights into the application of this LCA methodology within the construction industry.

The construction sector stands as a significant consumer of both renewable and non-renewable natural resources, occupying a prominent position in discussions surrounding environmental sustainability. As previously mentioned, the procedures and activities involved in construction demand substantial quantities of materials throughout its service life cycle. Therefore, the selection and utilization of eco-friendly construction materials assume paramount importance in the development and implementation of sustainable building practices [1].

Present Address:
M. A. Sánchez-Burgos (✉) · E. Conradi-Galnares
Department of Building Construction I, Higher Technical School of Architecture, University of Seville, Seville, Spain
e-mail: msanchez2@us.es

E. Conradi-Galnares
e-mail: conradi@us.es

Present Address:
D. Dragatogiannis
Delta Materials Process and Innovation Solutions, Agia Paraskevi, Greece
e-mail: ddragat@delta-ms.gr

© The Author(s) 2025

P. Mercader-Moyano and P. Porras-Pereira (eds.), *Life Cycle Analysis Based on Nanoparticles Applied to the Construction Industry*,
https://doi.org/10.1007/978-3-031-79115-4_3

1 Methodology and Phases of a Life Cycle Assessment

LCA addresses the environmental aspects and potential environmental impacts throughout a product's life cycle, spanning from the acquisition of raw materials, through production, usage, end-of-life treatment, recycling, and ultimate disposal [2].

In order to comprehensively assess the environmental burdens associated with processes and products throughout their entire life cycle, four essential steps must be undertaken, facilitating the comparison of different studies and ensure a holistic understanding of the environmental implications: goal and scope, Life Cycle Inventory (LCI), Life Cycle Impact Assessment (LCIA) and an interpretation (Fig. 1) [3].

- Scope and Goal definition

The first step, goal and scope, defines purpose, objectives, functional unit and system boundaries [5]. Defining the goal & scope of an LCA means defining what we want to analyse, how we want to analyse it, deciding which Impact Categories are going to be the focus of the assessment, and how far we want to go with our analysis [4].

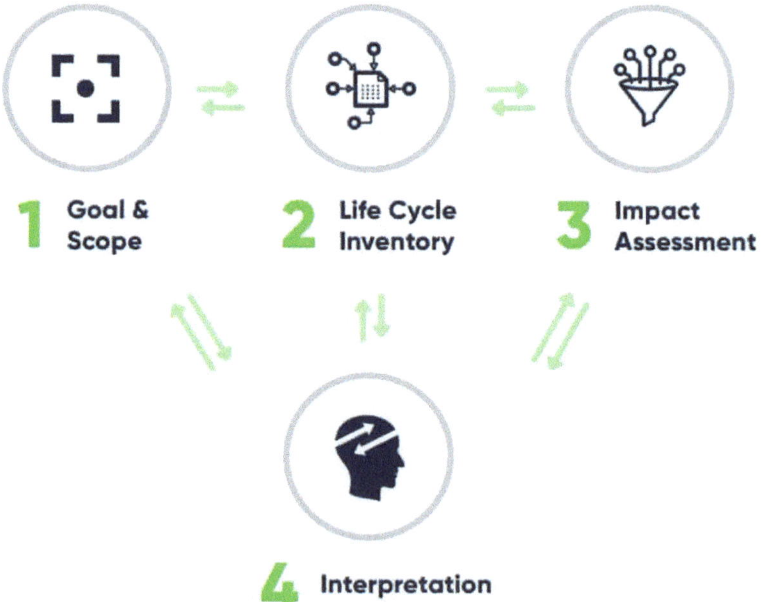

Fig. 1 Phases of a life cycle assessment [4]

One of the strengths of LCA is defining investigated products and processes based on their function instead of on their specific physical characteristics. This way, products can be compared that are inherently different, but fulfil a similar function [3].

- Lifecycle inventory (LCI) analysis

The subsequent phase, known as Life Cycle Inventory (LCI), involves the comprehensive gathering, description, and validation of data pertaining to inputs, processes, emissions, and other relevant factors across the entirety of the life cycle [3, 4]. Initially, this entails delineating a work breakdown structure to identify various components and subsequently computing the necessary quantities of materials required [6].

- Lifecycle impact assessment (LCIA)

Thirdly (LCIA), environmental impacts and used resources are quantified, based on the inventory analysis. This step contains three mandatory parts: precise selection of impact categories depending on the parameters of goal and scope, assignment of LCI results to the selected impact categories (classification) and calculation of category indicators (characterization) [3].

An impact category consolidates various emissions into a single environmental effect. Within the Life Cycle Impact Assessment (LCIA) process of a LCA, efforts are made to quantify these diverse emissions into tangible metrics. This involves aggregating emissions with similar environmental repercussions into unified units, facilitating the translation of disparate impacts into distinct impact categories [7].

Within the LCIA phase, two optional steps may be included: normalization and weighting. Normalization entails determining the scale of category indicator outcomes in relation to a reference dataset, such as the average environmental impact of a European citizen over the course of one year. Weighting involves the transformation of indicator outcomes from various impact categories into broader, overarching concerns or a singular score. This is achieved through the application of numerical factors derived from value-based decisions, such as policy objectives, monetization, or expert panel assessments. It represents the initial and significant stage in a LCA where subjective evaluations are introduced [3].

- Interpretation

The fourth and final step is the interpretation of the results, making the most reliable conclusions and recommendations [2, 3]. According to ISO 14044:2006, this is what the interpretation of a Life Cycle Assessment should include [4]:

- Identifying significant issues based on our LCI and LCIA phase.
- Evaluating the study itself, how complete it is, if it's done sensitively and consistently.
- Conclusions, limitations, and recommendations.

To gain a comprehensive understanding of the results, it's imperative to contextualize them within the broader framework of the analysis. This involves ensuring the collection of accurate data and meticulous attention to measurement and analysis procedures. By meticulously examining the data and methodologies employed, it can be ensured that the results are interpreted correctly and provide an accurate portrayal of the overall picture [4].

Figure 2 illustrates the different phases of a LCA, defining what is going to be analysed, how is going to be analysed, and how far the analysis will go in each stage [4].

The approaches to calculate environmental impacts can be classified into two fundamental groups: attributional and consequential LCA. Attributional LCA is characterized by its emphasis on delineating environmentally significant flows within a specified timeframe, while consequential LCA aims to describe how these flows might evolve in response to potential decisions [8]. Generally, most authors state that consequential LCAs are more appropriate for decision-making, unless their uncertainties in the modelling outweigh the insights gained from it [3]

One of the challenges in current LCA practices is the potential for disparate results when different methods are applied to the same case. For instance, a narrow-focused carbon footprint analysis may yield different outcomes compared to studies employing a broader set of impact indicators. Variations in methodology can attribute different levels of significance to various properties or impacts, consequently

Fig. 2 Development of the different phases of a Life Cycle Assessment [4]

suggesting divergent strategies for reducing ecological burdens. It's crucial to recognize that LCA results lack absolute values and thus cannot independently certify the sustainability of a product or service. However, they are invaluable for facilitating comparisons between different products and processes. Meaningful comparison between LCAs is contingent upon the subjects fulfilling identical functions as defined by their respective goals and scopes [3].

Another limitation lies in the incapacity to fully examine local impacts, as conventional LCA methodologies typically assess environmental damage on a global scale. However, this approach may overlook the nuances of localized conditions, where emissions and other factors can have heightened consequences, particularly in vulnerable regions. A more effective approach involves integrating LCA with specialized tools designed to evaluate local impacts, such as Risk Assessment methodologies [9]. This combination offers a more comprehensive understanding of environmental effects, allowing for better-informed decision-making regarding sustainability measures. Furthermore, local emissions can trigger additional repercussions, such as influencing the indoor climate of a building. While these emissions might not significantly impact the environment from a broader perspective, ensuring a healthy indoor environment necessitates the integration of supplementary criteria within the functional unit of an LCA or other localized damage assessments. Often, these criteria are included to adhere to regulatory standards. This underscores the challenge and significance of incorporating qualitative considerations into LCA methodologies, emphasizing the need for a more holistic approach to environmental assessment [3].

2 Impact Categories in LCA

As it has been defined before, an impact category can be understood as a class representing environmental issues of concern to which life cycle inventory analysis results may be assigned, it groups different emissions into one effect on the environment. Impact categories are helpful when make actionable statements about how Greenhouse Gas (GHG) emissions influence the environment.

Emissions manifest in various forms and compositions, reflecting the stark disparity between emissions generated during raw material harvesting and those generated in the process of electricity generation. This is where impact categories come into place. During the Life Cycle Impact Assessment (LCIA) of an LCA, these different emissions are unified into actionable numbers: different emissions that cause the same impact are converted into one unit that translates into one impact category [7].

Table 1 provides a comprehensive summary of 15 environmental impact categories, including their respective units of measurement and descriptions [7].

Table 1 Environmental impact categories [7]

Impact category/Indicator	Unit	Description
Climate change—total, fossil, biogenic and land use and land use change	kg CO_2-eq	Indicator of potential global warming due to emissions of greenhouse gases to the air. Divided into 3 subcategories based on the emission source: (1) fossil resources, (2) bio-based resources and (3) land use and land use change
Ozone depletion	kg CFC-11-eq	Indicator of emissions to air that causes the destruction of the stratospheric ozone layer
Acidification	kg mol H +	Indicator of the potential acidification of soils and water due to the release of gases such as nitrogen oxides and sulphur oxides
Eutrophication aquatic freshwater	kg P-eq	Indicator of the enrichment of the freshwater ecosystem with nutritional elements, due to the emission of nitrogen or phosphor-containing compounds
Eutrophication aquatic marine	Kg N-eq	Indicator of the enrichment of the marine ecosystem with nutritional elements, due to the emission of nitrogen-containing compounds
Eutrophication—terrestrial	mol N-eq	Indicator of the enrichment of the terrestrial ecosystem with nutritional elements, due to the emission of nitrogen-containing compounds
Photochemical ozone formation	kg NMVOC-eq	Indicators of emissions of gases that affect the creation of photochemical ozone in the lower atmosphere (smog) catalysed by sunlight
Depletion of abiotic resources—minerals and metals	kg Sb-eq	Indicator of the depletion of natural non-fossil resources
Depletion of abiotic resources—fossil fuels	MJ, net calorific value	Indicator of the depletion of natural fossil fuel resources
Human toxicity—cancer, non-cancer	CTUh	Impact on humans of toxic substances emitted to the environment. Divided into non-cancer and cancer-related toxic substances
Eco-toxicity (freshwater)	CTUe	Impact on freshwater organisms of toxic substances emitted to the environment
Water use	m^3 world eq. deprived	Indicator of the relative amount of water used, based on regionalized water scarcity factors
Land use	Dimensionless	Measure of the changes in soil quality (biotic production, erosion resistance, mechanical filtration)

(continued)

Table 1   (continued)

Impact category/Indicator	Unit	Description
Ionising radiation, human health	kBq U-235	Damage to human health and ecosystems linked to the emissions of radionuclides
Particulate matter emissions	Disease incidence	Indicator of the potential incidence of disease due to particulate matter emissions

3   Tools for the Development of LCA: Databases and Programs

Life cycle assessments are complex processes requiring a great quantity of input data and analysis [10]. Since environmental management standards UNE-EN ISO 14040 and UNE-EN ISO 14044 are in place in 2006, many LCA databases have been developed to assist in this process and make it easier [11]. Each has a specific goal (stated or implicit), such as reducing greenhouse gas emissions or improving indoor environmental quality [10].

Numerous databases are available for studying the environmental impact of construction materials throughout the building process; however, upon reviewing studies in this field, several significant issues have been identified. These include discrepancies between the locations covered by the LCA database and the actual study locations [11], a lack of transparency [12], and/or the unsuitability of the data for the specific conditions of the building project [13].

Initially, databases were academic niche projects. However, as LCA emerged as a pivotal instrument for driving environmental innovation, national governments eagerly adopted its methodologies. Nowadays, most LCA databases are developed by collaborations between governments, research groups, and national universities located in specific countries or territories, and model processes based on its regional manufacturing technologies [10]. In some cases, these databases do not even provide users with original data sources [14]. Additionally, many LCA databases predominantly employ a cradle-to-gate model, focusing solely on analysing the initial stages of a building's life cycle; however, this approach often overlooks other critical phases. Nevertheless, this omission presents an opportunity to incorporate additional factors, such as material transportation to the construction site and the management of construction and demolition waste (CDW), without duplicating efforts, provided that the transparency of the database allows for such modifications [15]. Despite the challenges highlighted earlier, significant advancements have been made in LCA methodology. Efforts to enhance database quality assurance, ensure methodological consistency, and harmonize approaches have all contributed to substantial progress in this field [16].

The European Commission´s Institute for Environment and Sustainability provides a compilation of available LCA databases.

- *Ecoinvent*

Ecoinvent was developed by the Swiss Centre for Life Cycle Inventories [17]. Due to its consistency and transparency, it has been included in SimaPro 7, and it is also possible to use it with GaBi 5 and Umberto 5. A cradle-to-gate model is applied in most of the LCA studies, which are based on downloadable reports [18–20]. In these reports, the methodology, flow charts, life cycle inventories, and literature references are presented. On paying the full licence, Ecoinvent is perfectly suited for construction, since every construction materials category is included and developed with a high variety of products. Users can also consult it online and download the data directly.

- *GaBi Database*

GaBi Database, by PE INTERNATIONAL, is one of the biggest LCA databases on the current market [21]. More than one thousand processes for construction materials are included, some of them from PlasticsEurope, ELCD, or Eurofer, and are cradle-to-gate. This is a complete database, which encompasses various processes, including those related to construction materials, with a diverse array of options within each product category. Moreover, the process data undergoes annual updates facilitated by a team of 50 LCA experts, ensuring its relevance and accuracy over time.

- *PlasticsEurope Eco-Profiles*

Eco-Profiles were first developed by PlasticsEurope in 1991 and have been continuously updated. This is a specialized and freely accessible LCA database focusing specifically on plastic materials. It offers cradle-to-gate data covering major polymers as they are manufactured in Europe. Notably, this database is integrated into both SimaPro and GaBi, widely used software platforms for conducting life cycle assessments. The methodology is complemented by intuitive flow charts and life cycle inventories for the different production stages. Eco-Profiles are calculated using average values from the industry, and then weighted depending on manufacturers' production. Raw material extraction, emissions to air and water, waste generation, as well as transportation related to production, vehicle maintenance, and the replacement of batteries and tires, are encompassed within the assessment. The energy and emissions linked to the construction of manufacturing plants, as well as human activities such as food consumption and commuting to work, are considered insignificant in comparison to industrial processes [22].

- *Athena database*

Athena database is used by Athena Impact Estimator. This database includes data for construction materials, energy, transport, construction and demolition processes, maintenance, repairing, and waste disposal, which is taken from U.S.LCI Database [23]. The data reflects the manufacturing process in Canada and the U.S.A, classified by regions, which considers differences in transport, energy mix and recycled material rates. All categories of construction materials are included, and it is capable of modelling 95% of the building stock in North America.

- *U.S. Life Cycle Inventory Database*

U.S. Life Cycle Inventory Database was developed in 2001 by the National Renewable Energy Laboratory of the U.S. Department of Energy. It was last updated in 2022 through a meeting with data providers. Data follows cradle-to-gate, cradle-to-grave and gate-to-gate models and considers input and output flows of energy and materials [24].

- *Base Carbone*

Created by the ADEME agency, and designed to determine greenhouse gas balances, the Base Carbone provides data on CO_2-emission factors for France and its islands and colonies. This freely accessible online database encompasses sets of metals, glass, plastics, and various other construction materials. It adopts a cradle-to-grave model, delineating life-cycle stages that encompass the manufacturing of new or recycled materials and their end-of-life phase. The output is quantified in equivalent kilograms of CO_2 per ton. Extensive documentation and reports accompany the data to provide comprehensive support. However, LCA inventories are somewhat limited, as this database primarily focuses on equivalent CO_2 emissions [25].

- *BEDEC database*

BEDEC is a structured database originating from Spain, developed by the Institute of Technology of Construction in Catalonia (IteC). It encompasses economic and environmental data for over half a million variations of construction elements. Specifically, the database offers information on embodied energy, CO_2 emissions, and waste disposal associated with each construction material. However, a notable drawback is the absence of accompanying documentation, which represents a significant inconvenience for its users [26]. The lack of traceability and outlining of the methodological process makes its data sets unverifiable. The data is from organizations and enterprises (e.g., Puerto de Barcelona, Grohe, Isover, Porcelanosa, Roca, Texsa), and works in collaboration with the ICAEN, the Polytechnic University of Catalonia and the Technological Centre of Construction. The data has been obtained from Ecoinvent, but no visible links are available to LCA studies in order to be consulted. The data on embodied energy of materials takes into account several factors, including raw material extraction, transportation from the source to the manufacturing facility, and the transformation processes carried out at the manufacturers. However, it does not encompass the transformation of these materials into construction products nor their transportation to the building site, then it can be defined as a cradle-to-gate model [26].

Despite all these problems, this is the most widely used database in Spain for environmental studies of a building construction process, since it matches very well with the variety of construction elements used, which makes it simple to use when the evaluator wishes to study an entire building [27].

- *CPM LCA Database*

CPM LCA Database, previously known as LCAiT, was developed in 1998 by the Swedish Center for Environmental Assessment of Product and Material Systems. This database is maintained by the University of Chalmers and contains manufacturing processes of construction materials. The database includes life cycle inventories, literature references and comprehensive documentation expressing the sources of each element of the studied process [28].

- *ProBas*

ProBas is a free online well-structured database, which can be read in GEMIS 4.8 [29]. Around seven hundred construction materials are included, belonging to all category types [30]. This comprehensive database includes references, life cycle inventories, and concise information for each entry. Most data come from the Öko-institut, whose headquarters are in Freiburg im Breisgau, Germany, complemented with LCA studies [31].

In Spain, there are not mandatory tools created by the administration that promote Green Public Procurement (GPP). There are private initiatives, such as HADES, which is a free tool which helps in the design for of more sustainable building. HADES is mainly oriented for housing, although it can also be used for offices and equipment. It evaluates energy, materials and circular economy, water, indoor environmental quality, and climate change. Other tool is VERDE also developed by GBCe which aims to provide a methodology for evaluating the sustainability of buildings. It is a methodology that allows us to conceive the building in a global way, including all its phases. VERDE develops tools for residential buildings, offices, rehabilitation, etc. LCA of the building is carried out where the potential impacts are weighted in absolute values and then these are associated to one of the 6 levels of certification through a comparison with a reference building. LEED y BREEAM, the first is also managed by GBCe and the second by BREEAM Spain, evaluate the CO_2 emissions caused by the manufacturing of materials employed and by the operational energy consumed. Its methodology is based on the determination of impacts classified by categories, although these impacts do not conform to the LCA. At the end of the process, like VERDE, it is the accredited certifier who calculates the rating of the building. Another tool available in Spain is ECOMETRO open source for the design, environmental assessment, and measurement of LCA in buildings. Among other aspects, it evaluates the carbon footprint in the construction and use phase, with the aim of obtaining zero CO_2 buildings.

These four tools (HADES, VERDE, BREEAM and ECOMETRO) have a life cycle approach, which requires specialized cognizance of environmental assessment procedures. Level(s) is a very ambitious but demanding proposal from the point of view of the previous knowledge needed to be able to use it correctly. In Spain, there is still room for improvement for easy-to-use tools, free, and that allow the self-evaluation by the user.

4 Environmental Certification Tools

Buildings exert significant direct and indirect impacts on the environment throughout their lifecycle stages, including construction, occupancy, renovation, repurposing, and demolition. These phases involve substantial consumption of energy, water, and raw materials, as well as the generation of waste and emissions with potential environmental consequences. In response to these challenges, various green building standards, certifications, and rating systems have been developed. These frameworks aim to address and mitigate the environmental impact of buildings by promoting sustainable design principles and practices [32].

Since the 1980s, green product standards have emerged in the marketplace, and their scope and prevalence have steadily increased. Initially, many of these standards arose in response to mounting apprehensions regarding product toxicity and its ramifications for children's health and indoor environmental quality (IEQ). In the present day, with escalating concerns regarding climate change and resource depletion backed by extensive research, the proliferation and diversity of green product standards and certifications persist. Furthermore, the focus of these standards has broadened to encompass a wider array of environmental considerations, including the entire lifecycle impacts of products from manufacturing to use and reuse. Despite the absence of a universal definition, green products are designed to fulfil claims of delivering environmental benefits while adhering to specific standards [32].

The marketplace is witnessing a proliferation of standards, rating systems, and certification programs aimed at guiding, demonstrating, and documenting endeavours to create sustainable, high-performance buildings. Globally, there exist hundreds of green product certifications, each employing diverse approaches. Some certifications delineate prerequisites alongside optional credits, offering flexibility in meeting sustainability goals. Others adopt a prescriptive approach, specifying detailed criteria for compliance. Additionally, certain certifications propose performance-based requirements that can be achieved through various methods tailored to specific products and project types [32].

Green building certification systems are increasingly recognized globally, with ongoing evolution to address the imperative of achieving net-zero emissions. These systems have established criteria for assessing sustainable actions within the building sector across various dimensions, including energy, water, waste, transportation, materials and resource consumption, pollution, land use, and human health. As a result, these certifications offer compelling evidence of the strides being made by the building sector towards embracing sustainable and eco-friendly practices in construction and development [32].

As of 2021, there were 74 green building certification systems worldwide, with the majority being administered by members of the World Green Building Council (WorldGBC). These certification systems have been utilized in at least 84 countries around the globe. However, despite the extensive range of available options, only 14 certification systems are included in the Global Buildings Climate Tracker, based on transparent and accessible data. Nonetheless, the number of certificates issued under these 14 certification systems exhibited robust growth, with an average cumulative increase of 19% in 2021. This marks a slight uptick compared to the 18% growth observed in 2020 and the 24% growth recorded in 2019 [32].

Green building certifications serve multiple purposes beyond environmental impact reduction. They contribute to the dissemination of local knowledge, fostering awareness and offering training opportunities in sustainable practices. Moreover, they serve as a crucial benchmark for sustainable investment and financing, functioning as a quality mark that enhances marketability and value. For instance, the European Bank for Reconstruction and Development mandates the use of green certifications such as EDGE, BREEAM, or LEED for all construction investments, underscoring their significance in the financing realm (Table 2). However, the limited recognition of local labels may hinder the adoption of locally recognized sustainability approaches. Establishing a clear framework for the utilization of certifications can facilitate collaboration among designers, investors, manufacturers, governmental and non-governmental organizations, thus driving collective efforts to accelerate decarbonization across the entire value chain [32].

Table 2 Global building certification programmes

Region	Country	Rating
Europe	Austria	DGNB Austria
	Croatia	DGNB Croatia
	Denmark	DGNB Denmark
	France	HQE
	Germany	DGNB
	Ireland	Home performance index
	Italy	Italy GBC Home, historic building, Quartieri, Condomini
	Latvia	BREEAM-LV
	Netherlands	BREEAM-NL, DGBC Woonmerk
	Norway	BREEAM-NOR
	Russia	OMIR
	Spain	DGNB Spain, VERDE
	Sweden	BREEAM-SE, Miljöbyggnad, Miljöbyggnad iDrift, CEEQUAL, NollCO$_2$
	Switzerland	Minergie, SNBS, DGNB Switzerland
	United Kingdom	BREEAM, EDGE

5 Life Cycle Assessment Conclusions

To mitigate the impact of material production on natural resources, it is imperative to advocate for the adoption of cutting-edge techniques and innovations in production facilities. Furthermore, wherever feasible, there should be a concerted effort to substitute finite natural resources with waste generated from various production processes, thereby closing the loop on product cycles. This necessitates a firm commitment to practices such as reuse and recycling, while simultaneously minimizing the transportation of raw materials and finished products. Encouraging the utilization of locally available resources further promotes sustainability by reducing reliance on distant sources and minimizing carbon emissions associated with transportation [5].

To enhance sustainability in the construction industry, it is crucial to extend, tailor, and standardize existing inventory databases of construction materials to align with the unique characteristics and requirements of each country's construction sector. Public institutions should play a proactive role in urging material manufacturers to adopt Environmental Product Declarations (EPDs) or Type III ecolabels, as per ISO standards. These EPDs, verified by independent entities, would provide standardized information based on the Life Cycle Assessment (LCA) of the real environmental impact of each product [5].

This approach would incentivize competition among manufacturers to introduce more eco-efficient products to the market. Such products, accompanied by EPDs, would be highly valued by the construction sector as they offer buildings with demonstrably low environmental footprints, not only in terms of energy consumption but also due to the reduced impact of the materials used [5].

Accurate information provided by EPDs would enable a more precise assessment of a building's environmental impact from an LCA perspective. Without this data, the assessment of a building's impact would rely on rough estimates derived from existing inventories, which may not adequately reflect the realities of a specific geographical area. Therefore, promoting the adoption of EPDs and standardized environmental labels is essential for advancing sustainability in the construction industry and ensuring informed decision-making [5]

The current practice of demolishing buildings at the end of their service life often results in the indiscriminate disposal of materials in landfills or incinerators, making recycling challenging. To enable effective recycling of construction materials, it is imperative to advocate for a fundamental shift in building design to facilitate the disassembly of materials at the end of their lifecycle [5].

To achieve this goal, buildings must be designed with a focus on ease of disassembly. This involves incorporating reversible joints between different materials, such as bolted joints, while minimizing the use of adhesives. This approach allows for the separation of materials without causing damage, facilitating their reuse, and recycling [5].

This paradigm shift in building design is essential to promote a circular economy in the construction sector. By prioritizing disassembly and recycling in building design, waste generation and resource depletion can be significantly reduced, advancing towards a more sustainable built environment [5].

In conclusion, the implementation of a sustainable building strategy should be integrated into a broader framework of sustainable development, aimed at achieving a reduction in overall resource consumption and exploitation. It is essential to avoid potential rebound effects and ensure a per capita decrease in the use of natural resources. To accomplish this objective, several measures should be considered. Firstly, moratoria should be established for the construction of new buildings and large infrastructures, promoting a more judicious use of land and resources [5].

By incorporating these aspects into a comprehensive sustainable decline strategy, a path towards achieving long-term environmental sustainability and resilience is paved, ensuring a balanced relationship between human activities and the natural world [5].

References

1. G. Ding, in *Life Cycle Assessment (LCA) of Sustainable Building Materials: An Overview*. Life Cycle Assessment (LCA), Eco-Labelling and Case Studies, pp. 38–62 (2014)
2. International Organization for Standardization (ISO), in *Environmental management—Life Cycle Assessment—Requirements and Guidelines* (ISO 14044:2006) [Online]. Available. https://www.iso.org/obp/ui/#iso:std:iso:14044:ed-1:v1:en. Accessed 09 Sept 2023
3. M. Buyle, J. Braet, A. Audenaert, Life cycle assessment in the construction sector: a review. Renew. Sustain. Energy Rev. **26**, 379–388 (2013)
4. Ecochain, in *Life Cycle Assessment (LCA)—Complete Beginner's Guide* [Online]. Available. https://ecochain.com/blog/life-cycle-assessment-lca-guide/. Accessed 06 Sept 2023
5. I. Zabalza Bribián, A. Valero Capilla, A. Aranda Usón, Life cycle assessment of building materials: comparative analysis of energy and environmental impacts and evaluation of the eco-efficiency improvement potential. Build. Environ. **46**(5), 1133–1140 (2011)
6. Y. Pamu, V. Kumar, M. Shakir, H. Ubbana, Life cycle assessment of a building using Open-LCA software. Mater. Today: Proc. **52**(3), 1968–1978 (2022)
7. Ecochain, Impact Categories (LCA)—Overview [Online]. Available: https://ecochain.com/blog/impact-categories-lca/. Accessed 05 Sept 2023
8. O. Ortiz-Rodríguez, F. Castells, G. Sonnemann, Life cycle assessment of two dwellings: One in Spain, a developed country, and one in Colombia, a country under development. Sci. Total. Environ. **408**(12), 2435–2443 (2010)
9. G. Finnveden, A. Moberg, Environmental systems analysis tools – an overview. J. Clean. Prod. **13**(12), 1165–1173 (2005)
10. Ecochain, in *LCI Databases in LCA: Function & Most-Used Databases*. [Online]. Available. https://ecochain.com/blog/lci-databases-in-lca/. Accessed 05 Sept 2023
11. I. Zabalza Bribián, A. Aranda Usón, S. Scarpellini, Life cycle assessment in buildings: State-of-the-art and simplified LCA methodology as a complement for building certification. Build. Environ. **44**(12), 2510–2520 (2009)
12. B. López-Mesa, in *Análisis de Ciclo de vida de Sistemas Constructivos/Life Cycle Assessment of Construction Systems*, Conference on Sustainability in Edification (2011)
13. A. Martínez-Rocamora, J. Solís-Guzmán, M. Marrero, LCA databases focused on construction materials: a review. Renew. Sustain. Energy Rev. **58**, 565–573 (2016)
14. ITeC, in *Construction cost bases. Institute of construction technology of Catalonia ITeC* (BEDEC) 2019. [Online]. Available: https://itec.es/banco-precios-bedec/. Accessed 06 Sept 2023

15. M. Marrero, C. Rivero-Camacho, M.D. Alba-Rodríguez, What are we discarding during the life cycle of a building? Case studies of social housing in Andalusia, Spain. Waste Manage. **102**, 391–403 (2020)
16. European Commission, European Platform on Life Cycle Assessment [Online]. Available: https://data.jrc.ec.europa.eu/collection/EPLCA. Accessed 20 May 2023
17. Ecoinvent Centre, in *Ecoinvent* [Online]. Available. https://ecoinvent.org/database/. Accessed 27 May 2023
18. H.C.M. Althaus, Life cycle inventories of metals and methodological aspects of inventorying material resources in ecoinvent (7 pp). J. Life Cycle Assess. **10**, 43–49 (2005)
19. D. Kellenberger, H. J. Althaus, T. Künniger, M. Lehmann, in *Life Cycle Inventories of Building Products* (Final Report Ecoinvent Data v2.0 No. 7, 2007) [Online]. Available: https://db.eco invent.org/reports/07_BuildingProducts.pdf?area=463ee7e58cbf8. Accessed 06 Sept 2023
20. R. Hischier, in *Life Cycle Inventories of Packaging and Graphical Papers.* (Ecoinvent-Report No. 11, 2007). [Online]. Accessed 27 May 2023
21. Greenhouse Gas Protocol, in *Construction Materials extension* (GaBi Database website) [Online]. Available. https://ghgprotocol.org/gabi-databases. Accessed 27 May 2023
22. PlasticsEurope, PlasticsEurope Eco-Profiles website [Online]. Available. https://plasticse urope.org/sustainability/circularity/life-cycle-thinking/eco-profiles-set/. Accessed 20 May 2023
23. Athena SMI, Athena Database Content [Online]. Available. https://www.calculatelca.com/wp-content/uploads/2012/10/LCI_Databases_Products.pdf. Accessed 20 May 2023
24. NREL (National Renewable Energy Laboratory), U.S. Life Cycle Inventory Database [Online]. Available. https://www.nrel.gov/lci/. Accessed 20 May 2023
25. ADEME, Base Carbone Website [Online]. Available. https://data.ademe.fr/datasets/base-car boner. Accessed 20 May 2023
26. ITeC, BEDEC Website [Online]. Available. https://itec.es/servicios/bedec/. Accessed 20 May 2023
27. P. Mercader-Moyano, Cuantificación de los recursos consumidos y emisiones de CO_2 producidas en las construcciones de Andalucía y sus implicaciones en el protocolo de Kioto, Sevilla, 2010.
28. CPM, CPM LCA Database Website [Online]. Available. http://cpmdatabase.cpm.chalmers.se/Start.asp. Accessed 28 May 2023
29. IINAS, GEMIS—Global Emissions Model for Integrated Systems Website [Online]. Available. http://www.iinas.org/gemis.html. Accessed 28 May 2023
30. ProBas, ProBas database website [Online]. Available. https://www.probas.umweltbundesamt.de/php/index.php. Accessed 28 May 2023
31. U. Hantsche, Abschätzung des kumulierten Energieaufwandes und der damit verbundenen Emissionen zur Herstellung ausgewählter Baumaterialien/Assessment of the cumulative energy consumption and related emissions for the production of selected construction materials, VDI-Berichte 1093, 1993.
32. S. Vierra, Green Building Standards and Certification Systems, 2023. [Online]. Available. https://www.wbdg.org/resources/green-building-standards-and-certification-systems#rcas. Accessed 19 Sept 2023

Part II
Nanoproducts in the Construction Sector

Chapter 4
Advanced Materials: Introduction to Nanotechnology

Begoña Blandón-González, Aikaterini Flora Trompeta, and Pilar Mercader-Moyano

Abstract The initial segment unfolds by expounding upon the foundational concepts of nanotechnology, tracing its historical evolution, delving into its normative framework of reference, and elucidating its diverse applications across various sectors, navigating the inherent challenges associated with nanotechnology.

1 Historical Approach to New Materials

The nanotechnology revolution is making a ground-breaking impact on diverse science, engineering, and commercial sectors, including the construction industry. The physical and chemical properties unique to the nanoscale can lead to remarkable efficacy enhancement in (photo)catalysis, (thermal and electrical) conductivity, mechanical strength, and optical sensitivity, enabling applications such as catalysts, electronic and energy storage devices, advanced mechanical materials, and sensors [1].

The earliest ideas surrounding nanotechnology can be traced back to the pioneering insights of physicist Richard Feynman (Nobel Prize in physics 1965), who, in his seminal lecture titled "There is a lot of room in the background", presented at the annual meeting of the American Physical Society in 1959. Within his discourse,

Present Address:
B. Blandón-González · P. Mercader-Moyano (✉)
Department of Building Construction I, Higher Technical School of Architecture, University of Seville, Seville, Spain
e-mail: pmm@us.es

B. Blandón-González
e-mail: bblandon@us.es

Present Address:
A. F. Trompeta
Department of Materials Science and Engineering, School of Chemical Engineering Iroon Polytechniou, National Technical University of Athens (NTUA), Athens, Greece
e-mail: ktrompeta@chemeng.ntua.gr

© The Author(s) 2025
P. Mercader-Moyano and P. Porras-Pereira (eds.), *Life Cycle Analysis Based on Nanoparticles Applied to the Construction Industry,*
https://doi.org/10.1007/978-3-031-79115-4_4

Feynman eloquently expressed: "Most cells are tiny, but they are very active, they make substances, they move, they contort, and they do a lot of wonderful things, all on a small scale. They also store information. Let's consider the possibility that we, too, could make entities so small that they did whatever we wanted, that we could make an object that maneuverer at that level" [2].

The word nanotechnology was first used by Norio Taniguchi in 1974, to designate a nanoscale production technique, which involves processes of separation, consolidation, and deformation of materials with the help of a single atom or a single molecule. But it was not really disseminated until 1986 by Eric Drexler in his book "Creation Machines", in which he described the bases for the construction of atom-by-atom materials that "would open the doors to a technological development unprecedented in the history of mankind, which would make it possible to overcome disease and death, make intergalactic trips and have almost infinite material resources" [2, 3].

Louis Brus, working at AT&T Bell Laboratories, discovered colloidal semiconductor nanocrystals, called quantum dots, that led him to be recognized as one of the leading researchers in the field of nanoscience in the early 1980s [2].

In 1985 Sir Harold W. Kroto, Richard E. Smalley and Robert F. Curl, Jr. (Nobel Prize in Chemistry 1996) obtained hollow carbon molecules that formed a closed cage by using a laser to vaporize graphite rods in a helium gas atmosphere. The capsule comprised 60 carbon atoms intricately bonded through both single and double bonds, meticulously arranged to fashion a hollow sphere characterized by 12 pentagonal and 20 hexagonal faces. Such nanostructures, exemplified by this configuration, are commonly referred to as fullerenes [2, 4].

Japanese physicist Sumio Iijima in 1991 discovered and detailed the atomic structure and helical shape of single- and multi-walled carbon nanotubes [2].

In 1993, at the Massachusetts Institute of Technology, Moungi Bawendi and his collaborators developed a synthesis of nanocrystals obtaining semiconductor colloidal quantum dots, this nanotechnology being one of the first to be integrated with the biological sciences [5].

Since 1998 chemical engineer Thomas Webster worked on the design, synthesis and evaluation of nanomaterials for various medical applications. This included self-assembled chemistries, nanoparticles, nanotubes and nanostructured surfaces. Some of the medical applications included inhibiting the growth of bacteria, controlling inflammation, and promoting tissue growth [6, 7].

In the new century multiple contributions have been generated, among them are the gold nanocapsules for cancer treatment of Medicine and Surgery; nanoscale assembled devices similar to DNA and a methodology to generate nanoscale patterns and structures [2, 8].

2   Nanotechnology: Basic Concepts

Nanotechnology has been a recognized field of inquiry since the previous century, notably propelled into the scientific Richard P. Feynman during his aforementioned seminal lecture "There's Plenty of Room at the Bottom" in 1959. Since then, the landscape of nanotechnology has been marked by a succession of groundbreaking advancements. This domain has witnessed the synthesis and manipulation of materials at the nanoscale level, ushering in a new era of scientific exploration and technological innovation [9].

Central to the realm of nanotechnology are nanoparticles (NPs), a diverse class of materials encompassing particulate substances characterized by at least one dimension measuring less than 100 nm. The significance of these nanostructures became evident as researchers uncovered their capacity to profoundly influence the physiochemical properties of materials. Indeed, the size of nanoparticles emerged as a critical factor, intricately shaping phenomena such as optical properties, and offering tantalizing prospects for tailored material functionalities [9].

As aforementioned, nanoparticles are complex entities, consisting of three distinct layers: (a) The surface layer, which can be functionalized with various small molecules, metal ions, surfactants, and polymers, (b) the shell layer, a chemically distinct material from the core in all aspects, and (c) the core, which is essentially the central portion of the NP and usually refers the NP itself [10]. Due to these remarkable attributes, researchers across diverse fields have been captivated by the potential of these materials, sparking widespread interest and enthusiasm among multidisciplinary practitioners [9].

NPs are extensively categorized into different classes, delineated by their morphology, dimensions, and chemical attributes. Notable among these classifications are the following well-recognized categories, distinguished based on their distinct physical and chemical characteristics [9]:

- Carbon-based NPs

Fullerenes and carbon nanotubes (CNTs) stand out as prominent categories within the realm of carbon-based nanoparticles. Fullerenes are characterized by their globular hollow cage-like structures, comprising various allotropic forms of carbon. These nanostructures have garnered substantial commercial attention owing to their exceptional electrical conductivity, impressive strength, unique structural attributes, high electron affinity, and remarkable versatility [11]. Given their distinctive physical, chemical, and mechanical properties, these materials find utility not only in their pristine form but also as integral components in nanocomposites, facilitating a myriad of commercial applications. These applications include utilization as fillers, efficient gas adsorbents for environmental remediation, and support mediums for a diverse array of inorganic and organic catalysts [9].

- Metal NPs

Metal nanoparticles are exclusively composed of metal precursors. Harnessing the intrinsic properties of these metals, metal NPs exhibit distinct optoelectrical behaviour attributed to their well-documented localized surface plasmon resonance (LSPR) phenomenon. Among the alkali and noble metals, including copper (Cu), silver (Ag), and gold (Au), these NPs display a broad absorption band within the visible range of the electromagnetic solar spectrum [12].

The controlled synthesis of metal NPs, with precise manipulation of their facets, sizes, and shapes, is paramount in contemporary materials science and pivotal in the development of cutting-edge materials, enabling tailored properties and functionalities that cater to diverse technological applications [12]. Metal nanoparticles, owing to their sophisticated optical properties, are instrumental across various research domains. Notably, gold nanoparticles (NPs) coatings are extensively employed in the realm of scanning electron microscopy (SEM), in order to augment the electronic beam, thereby enhancing the signal-to-noise ratio and overall quality of SEM images [9].

- Ceramics NPs

Ceramic nanoparticles represent a class of inorganic non-metallic solids, synthesized through processes involving intense heat followed by controlled cooling. Exhibiting a diverse array of morphologies including amorphous, polycrystalline, dense, porous, or hollow structures, these nanoparticles have garnered significant interest from researchers. Their versatile properties render them indispensable in various applications such as catalysis, photocatalysis, the photodegradation of dyes, and imaging applications [9, 13].

- Semiconductor NPs

Semiconductor materials occupy a unique position between metals and nonmetals, endowing them with a spectrum of properties that make them highly versatile in various applications [14]. Semiconductor nanoparticles, in particular, are characterized by wide bandgaps, which render them susceptible to significant alterations in their properties through bandgap tuning. As a result, these nanoparticles play a pivotal role in fields such as photocatalysis, photo optics, and electronic devices [9].

For instance, semiconductor nanoparticles exhibit remarkable efficiency in water splitting applications, attributed to their suitable bandgap and bandedge positions. This capability to efficiently harness solar energy for catalytic reactions underscores the pivotal role of semiconductor nanoparticles in advancing sustainable energy technologies and addressing environmental challenges [9].

- Polymeric NPs

Typically derived from organic materials, nanoparticles in the literature are often referred to collectively as polymer nanoparticles (PNPs). These nanoparticles predominantly exhibit nanospheric or nanocapsular shapes [15]. Nanospheres represent matrix particles with a solid overall mass, with additional molecules adsorbed at the outer boundary of their spherical surface. Conversely, nanocapsules encapsulate a solid mass entirely within the particle structure [9, 16].

One of the distinguishing characteristics of PNPs is their high degree of functionality, enabling researchers to modify their surface properties with ease. This attribute renders PNPs highly versatile and applicable across a wide range of fields in scientific literature [9, 16].

- Lipid-based NPs

Nanoparticles containing lipid moieties have emerged as valuable assets in numerous biomedical applications. Characteristically spherical, lipid nanoparticles typically exhibit diameters ranging from 10 to 1000 nm. Similar to polymeric nanoparticles, lipid nanoparticles feature a solid core composed of lipids, while a matrix within them accommodates soluble lipophilic molecules. The external core of these nanoparticles is stabilized by surfactants or emulsifiers [9, 17].

Lipid nanotechnology represents a specialized field dedicated to the design and synthesis of lipid nanoparticles for diverse applications, such as drug delivery and carrier systems, as well as the targeted release of RNA in cancer therapy [9, 17].

Nanomaterials are further categorized based on their origins into natural, incidental, and artificial types. Natural nanomaterials originate from sources such as trees, plants, volcanoes, or marine species. Incidental nanomaterials are generated during processes like combustion in vehicles and industrial activities. Artificial nanomaterials, the most prevalent category, are produced through two primary manufacturing processes: top-down and bottom-up techniques (Fig. 1) [2].

Top-down techniques involve the reduction of macroscopic materials or solid groups to nanoscale dimensions. Physical methods like grinding or abrasion, chemical processes, and the vaporization of solids followed by condensation are employed

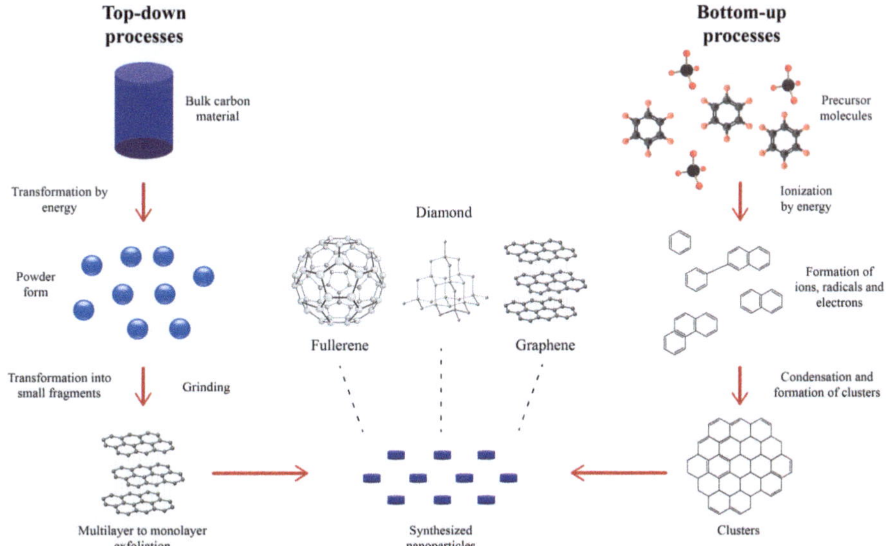

Fig. 1 Nanoparticle synthesis techniques [2]

to achieve controlled assemblies until the desired size is reached. Conversely, bottom-up techniques involve the creation of nanoparticles capable of self-assembly or self-organization through the condensation of atoms or molecular entities in a gaseous phase or solution. These techniques facilitate the precise construction of nanoparticles from atomic or molecular components, enabling tailored design and functionality [2].

3 Nanotechnology Applications in Different Sectors: Challenges and Trends

Considering their unique properties, NPs can be used in a spectrum of applications. Some significant examples of their versatility and practical use are given below [9]

- Applications in drugs and medications

Nano-sized inorganic particles, whether simple or intricate in nature, exhibit unparalleled physical and chemical characteristics, rendering them indispensable in the advancement of innovative nanodevices. These nanomaterials play a pivotal role across diverse realms including physical sciences, biology, biomedicine, and pharmaceuticals [1].

NPs have garnered escalating attention across various medical disciplines due to their capacity to deliver drugs within the optimal dosage range. This often translates into heightened therapeutic efficacy, diminished side effects, and enhanced patient adherence. When selecting NPs for applications such as biological and cellular imaging, as well as photothermal therapy, emphasis is placed on their optical properties. The ongoing pursuit of hydrophilic NPs as drug carriers reflects a significant challenge in recent years, signifying a pivotal area of research and development in pharmaceutical science [9].

Superparamagnetic iron oxide NPs, when endowed with suitable surface chemistry, boast a wide array of in vivo applications, such as MRI contrast enhancement, tissue repair, and immunoassay, detoxification of biological fluids hyperthermia, drugs delivery and cell separation [9].

In recent decades, there has been a notable surge in the exploration of biodegradable nanoparticles (NPs) as potent drug delivery systems. Researchers have leveraged a plethora of polymers in this pursuit, capitalizing on their ability to efficiently transport drugs to targeted sites. This strategy enhances therapeutic efficacy while mitigating potential side effects. Central to this endeavour is the concept of controlled drug release, aiming to administer pharmacologically active agents precisely to the desired site of action, at the optimal dosage and frequency [9].

Semiconductor and metallic nanoparticles stand out for their significant potential in both cancer diagnosis and therapy, owing to their ability to leverage surface plasmon resonance (SPR) for enhanced light scattering and absorption. Silver nanoparticles (Ag NPs) are witnessing growing utilization in wound dressings, catheters, and a variety of household products, thanks to their remarkable antimicrobial properties. The presence of such antimicrobial agents is critical in diverse sectors such as textiles, medicine, water disinfection, and food packaging, where they play a vital role in maintaining hygiene and preventing the spread of infectious agents [9].

- Applications in manufacturing and materials

Among consumer products incorporating nanotechnology, health and fitness products dominate the market, constituting the largest category. They are closely followed by products in the electronic and computer sector, as well as those in the home and garden category. Nanotechnology has garnered considerable attention as the next revolutionary force across numerous industries, with anticipated transformations in areas such as food processing and packaging [9].

Resonant energy transfer (RET) systems, comprising organic dye molecules and noble metal nanoparticles, have emerged as a focal point of interest in both bio photonics and material science. The integration of NPs into commercially available products is increasingly prevalent, reflecting the growing recognition of their potential across various applications [9, 18].

Metal nanoparticles, particularly noble metals like gold (Au) and silver (Ag), exhibit a diverse array of colours in the visible spectrum owing to their plasmon resonance phenomenon. This resonance arises from the collective oscillations of electrons at the nanoparticle surface [19]. The wavelength of this resonance is highly influenced by factors such as nanoparticle size and shape, interparticle distance, and the dielectric properties of the surrounding medium. Leveraging these unique plasmon absorbance characteristics, noble metal nanoparticles have found extensive utility in various applications, notably including chemical and biosensors [19].

- Applications in the environment

The growing utilization of engineered nanoparticles in both industrial and household settings raise concerns regarding their release into the environment. Evaluating the environmental risk posed by these NPs necessitates a comprehensive understanding of their mobility, reactivity, ecotoxicity, and persistence. Of particular concern is the potential for increased NP concentrations in groundwater and soil due to engineering material applications, as these represent the primary exposure pathways for assessing environmental risks [9].

Natural nanoparticles play a significant role in the partitioning of contaminants between solid and water phases due to their high surface-to-mass ratio. Contaminants can adhere to the surface of NPs, become co-precipitated during the formation of natural NPs, or be trapped through NP aggregation, where contaminants are initially adsorbed. The interaction between contaminants and NPs is influenced by various characteristics of the NPs, including their size, composition, morphology, porosity, aggregation/disaggregation tendencies, and aggregate structure. Additionally, in environmental contexts, luminophores, while not inherently safe, can be shielded from environmental oxygen by encapsulating them within the silica network [9].

The majority of environmental applications of nanotechnology can be classified into three main categories:

1. Environmentally benign sustainable products.
2. Remediation of materials contaminated with hazardous substances.
3. Sensors for environmental stages.

The extraction of heavy metals, such as mercury, lead, thallium, cadmium, and arsenic, from natural water sources has garnered significant interest due to their detrimental impacts on both environmental ecosystems and human health. Superparamagnetic iron oxide nanoparticles have emerged as a potent sorbent material for efficiently capturing these hazardous elements from water, presenting a promising solution for addressing the challenges posed by such toxic contaminants [9].

The scarcity of measurements for engineered nanoparticles in the environment can be attributed to the absence of analytical techniques capable of precisely quantifying trace concentrations of NPs. Additionally, it is noteworthy that photodegradation by NPs is a prevalent phenomenon, with many nanomaterials being deliberately employed for this purpose [9].

- Applications in electronics

In recent years, there has been a burgeoning interest in the advancement of printed electronics. This interest stems from the allure of printed electronics as a compelling alternative to conventional silicon-based techniques, which hold the promise of delivering low-cost, large-area electronic solutions, particularly well-suited for flexible displays and sensor applications [20]. The rapid adoption of printed electronics, which incorporate diverse functional inks containing nanoparticles such as metallic NPs, organic electronic molecules, CNTs, and ceramic NPs, is anticipated to revolutionize mass production processes for novel electronic devices [20].

The distinctive structural, optical, and electrical characteristics exhibited by one-dimensional semiconductors and metals position them as pivotal building blocks for the development of a new generation of electronic, sensor, and photonic materials [21].

A prime illustration of the synergy between scientific discovery and technological advancement is exemplified in the electronic industry. The exploration and identification of novel semiconducting materials have sparked a transformative revolution, transitioning from vacuum tubes to diodes and transistors, ultimately culminating in the miniaturization of electronic components into integrated circuits [9].

The important attributes of nanoparticles include their inherent ease of manipulation and reversible assembly, enabling their integration into electric, electronic, or optical devices. This capability, exemplified by "bottom-up" or "self-assembly" approaches, serves as a hallmark of nanotechnology, offering versatile pathways for the design and fabrication of advanced nanoscale devices [9].

- Applications in energy harvesting

Recent research has raised concerns regarding the limitations and scarcity of fossil fuel reserves in the foreseeable future, attributed to their non-renewable nature.

Consequently, scientists are shifting their research focus towards developing renewable energy sources derived from readily available and cost-effective resources. Nanoparticles have emerged as prime candidates for this purpose due to their expansive surface area, optical properties, and catalytic capabilities. Particularly in photocatalytic applications, NPs play a significant role in harnessing energy through photoelectrochemical (PEC) and electrochemical water splitting [22, 23]. Furthermore, NPs are extensively utilized in electrochemical CO_2 reduction to produce fuel precursors, as well as in the fabrication of solar cells and piezoelectric generators, offering promising avenues for energy generation [9, 24–26]. Additionally, NPs find applications in energy storage, enabling the storage of energy in various forms at the nanoscale level [9].

Innovatively, nanogenerators have been developed to convert mechanical energy into electricity, utilizing piezoelectric principles, which is an unconventional approach that represents a significant advancement in energy generation technology [27].

- Applications in mechanical industries

The remarkable mechanical properties of nanoparticles, including their exceptional Young's modulus, stress, and strain characteristics, unlock numerous applications in mechanical industries, particularly in coatings, lubricants, and adhesive applications. Moreover, these properties facilitate the development of mechanically robust nanodevices for diverse purposes. By embedding nanoparticles in metal and polymer matrices, tribological properties can be finely tuned at the nanoscale level, resulting in enhanced mechanical strength [9].

Furthermore, nanoparticles exhibit favourable sliding and delamination properties, contributing to reduced friction and wear, thus enhancing lubrication effects. Coatings incorporating nanoparticles can impart a range of robust mechanical characteristics, enhancing toughness and wear resistance. Notably, alumina, titania, and carbon-based nanoparticles have proven successful in achieving desirable mechanical properties in coatings [9, 28].

4 Normative Frame of Reference for Nanoproducts in Construction Sector

In the European Union, nanomaterials are subject to the same rigorous regulatory framework as other chemicals and mixtures, namely the Registration, Evaluation, Authorization, and Restriction of Chemicals (REACH) and Classification, Labeling, and Packaging (CLP) regulations. This mandates the assessment of hazardous properties specific to nanoforms of substances and ensures the safe use of nanomaterials [29].

To define the term nanomaterials, the European Commission has issued a recommendation focused on the size of constituent particles within a material, regardless of hazard or risk considerations. This definition encompasses materials of natural, incidental, or manufactured origin and serves as the foundation for regulatory measures governing this class of materials. However, in certain legislative domains, the driver for legal obligations or nanomaterials stem from their potential to exhibit different properties compared to larger particles [29].

For legal manufacturing or importation within the EU, all substances falling under the scope of REACH must undergo registration. This process mandates that manufacturers and/or importers provide information on both human health and environmental impacts, encompassing any hazardous nanoforms. Additionally, they are required to estimate exposure levels throughout the life cycle of the substance [29]. The same obligations extend to nanomaterials within the EU. When nanomaterials exhibit hazardous properties, they must be notified to the ECHA and appropriately labelled and packaged in accordance with the CLP regulation. This ensures that nanomaterials can be safely used in compliance with regulatory standards [29].

It is imperative for companies to be transparent in their REACH registration process, particularly regarding clearly addressing how the safety of nanoforms has been managed. This includes delineating the necessary measures to effectively control potential risks associated with these nanoforms. To assist companies in fulfilling these obligations, the ECHA offers guidance documents that provide further support to companies on how to identify and report properties of nanoforms, aiding in the comprehensive and accurate registration of nanomaterials under REACH [29].

As of 1 January 2020, companies engaged in the manufacturing or importing of nanoforms are subject to explicit legal requirements under REACH. These reporting obligations address specific information requirements as outlined in revised annexes to the REACH regulation [30]:

- characterisation of nanoforms or sets of nanoforms covered by the registration (Annex VI) [30].
- chemical safety assessment (Annex I) [30].
- registration information requirements (Annexes III and VII-XI) [30].
- downstream user obligations (Annex XII) [30].

The amendments encompass both new and existing registrations involving nanoforms [30].

Given that nanomaterials fall within the scope of REACH and CLP regulations, it is essential for the ECHA to effectively execute its responsibilities concerning nanoforms within the various REACH (e.g., registration, evaluation, authorization, and restrictions) and CLP processes (e.g., classification and labelling) as it would for any other form of a substance. To achieve this, ECHA must possess adequate scientific and technical capacity [30].

With this objective in mind, ECHA has increased its efforts in this domain since 2011, with a concentrated focus on [30]:

- Developing new and updated guidance documents [30].
- Enhancing both internal and external capacity building [30].

- Facilitating dialogue and fostering consensus among Member State Competent Authorities, as well as members of the risk assessment and Member State committees, regarding safety information pertaining to nanomaterials included in REACH registration dossiers [30].
- Offering feedback and guidance to companies engaging in the registration of nanomaterials [30].
- Engaging actively and contributing to ongoing international regulatory initiatives, such as the OECD Working Party on Manufactured Nanomaterials or the Malta Initiative aimed at developing test guidelines [30].
- Conducting webinars to disseminate information and facilitate discussions on recent advancements concerning REACH and CLP processes related to nanomaterials. These sessions aim to assist registrants in preparing and submitting dossiers that involve nanomaterials, ensuring compliance with regulatory requirements [30].
- The Nanomaterials Expert Group (NMEG) was established in October 2012 with the backing of the competent authorities for REACH and CLP (CARACAL) and for Biocides. This informal advisory group plays a pivotal role in bolstering the execution of ECHA's workplans for nanomaterials. It furnishes guidance and counsel on scientific and technical matters pertaining to the implementation of REACH, CLP, and BPR legislation concerning nanomaterials [30].
- Hosting the European Union Observatory for Nanomaterials as a means to enhance the transparency of information pertaining to nanomaterials [30].

In the European Union, legislation concerning environmental, worker, and consumer protection is commonly implemented through directives. In cases where nanomaterials present potential risks to the environment, workers, or consumers, the overarching regulations established within the legislation are applicable to nanomaterials in the same manner as they are to other forms of the substance [29].

References.

1. J. Lee, S. Mahendra, P.J.J. Alvarez, Nanomaterials in the construction industry: a review of their applications and environmental health and safety considerations. ACS Nano **4**(7), 3580–3590 (2010)
2. M. Gómez Garzón, Nanomateriales, Nanopartículas y Síntesis Verde, Revista Repertorio de Medicina y Cirugía. Fundación Universitaria de Ciencias de la Salud—FUCS (2018)
3. J. Kreuter, Nanoparticles—a historical perspective. Int. J. Pharm. **331**(1), 1–10 (2007)
4. N. Martín, Sobre fullerenos, nanotubos de carbono y grafenos, Arbor **187**(Extra_1_), 115–131 (2011)
5. A. Valizadeh, H. Mikaeili, M. Samiei, S. Farkhani, N. Zarghami, M. Kouhi, A. Akbarzadeh, S. Davaran, Quantum dots: synthesis, bioapplications, and toxicity. Nano Rev. **7**(480) (2012)
6. I. Seil, T. Webster, Antimicrobial applications of nanotechnology: methods and literature. Int. J. Nanomed. **7**, 2767–3278 (2012)
7. E. Taylor, T. Webster, Reducing infections through nanotechnology and nanoparticles. Int. J. Nanomed. **6**, 1463–1473 (2011)

8. F.J. Heiligtag, M. Niederberger, The fascinating world of nanoparticle research. Mater. Today **16**(7–8), 262–271 (2013)

9. I. Khan, K. Saeed, Nanoparticles: properties, applications and toxicities. Arab. J. Chem. **12**(7), 908–931 (2019)

10. W. Shin, J. Cho, A. Kannan, Y. Lee, D. Kim, Cross-linked composite gel polymer electrolyte using mesoporous methacrylate-functionalized SiO_2 nanoparticles for lithium-ion polymer batteries. Sci. Rep. **6**(26332) (2016)

11. A. Astefanei, O. Núñez, M. Galceran, Characterisation and determination of fullerenes: a critical review. Anal. Chim. Acta **882**(2), 1–21 (2015)

12. E. Dreaden, A. Alkilany, X. Huang, C. Murphy, M. El-Sayed, The golden age: gold nanoparticles for biomedicine. Chem. Soc. Rev. **41**(7), 2740–2779 (2012)

13. S. Thomas, Harshita, P. Mishra, S. Talegaonkar, Ceramic nanoparticles: fabrication methods and applications in drug delivery. Curr. Pharm. Des. **21**(42), 6165–6188 (2015)

14. S. Ali, I. Khan, S. Khan, M. Sohail, R. Ahmed, A. Rehman, M. Shahid Ansari, M. Ali Morsy Electrocatalytic performance of Ni@Pt core–shell nanoparticles supported on carbon nanotubes for methanol oxidation reaction. J. Electroanal. Chem. **795**, 17–25 (2017)

15. M. Mansha, I. Khan, N. Ullah, A. Qurashi, Synthesis, characterization and visible-light-driven photoelectrochemical hydrogen evolution reaction of carbazole-containing conjugated polymers. Int. J. Hydrog. Energy **42**(16), 10952–10961 (2017)

16. N.H. Abd Ellah, S.A. Abouelmagd, Surface functionalization of polymeric nanoparticles for tumor drug delivery: approaches and challenges. Expert. Opin. Drug Deliv. **14**(2), 201–214 (2017)

17. M. Gujrati, A. Malamas, T. Shin, E. Jin, Y. Sun, Z. Lu, Multifunctional cationic lipid-based nanoparticles facilitate endosomal escape and reduction-triggered cytosolic siRNA release. Mol. Pharm. **11**(8), 2734–2744 (2014)

18. Y. Lei, W. Huang, M. Zhao, Y. Chai, Y. R. and Y. Zhuo, "Electrochemiluminescence resonance energy transfer system: mechanism and application in ratiometric aptasensor for lead ion. Anal. Chem. **87**(15), 7787–7794 (2015)

19. S. Unser, I. Bruzas, J. He, L. Sagle, Localized surface plasmon resonance biosensing: current challenges and approaches. Sensors **15**(7), 15684–15716 (2015)

20. A. Kosmala, R. Wright, Q. Zhang, P. Kirby, Synthesis of silver nano particles and fabrication of aqueous Ag inks for inkjet printing. Mater. Chem. Phys. **129**(3), 1075–1080 (2011)

21. M. Shaalan, M. Saleh, M. El-Mahdy, M. El-Matbouli, Recent progress in applications of nanoparticles in fish medicine: a review. Nanomed. Nanotechnol. Biol. Med. **12**(3), 701–710 (2016)

22. F. Ning, M. Shao, S. Xu, Y. Fu, R. Zhang, M. Wei, D. Evans, X. Duan, TiO_2/graphene/NiFe-layered double hydroxide nanorod array photoanodes for efficient photoelectrochemical water splitting. Energy Environ. Sci. **9**, 2633–2643 (2016)

23. V. Avasare, Z. Zhang, D. Avasare, I. Khan, A. Qurashi, Room-temperature synthesis of TiO_2 nanospheres and their solar driven photoelectrochemical hydrogen production. Energy Res. **39**(12), 1714–1719 (2015)

24. M. Gawande, A. Goswami, F. Felpin, T. Asefa, X. Huang, R. Silva, X. Zou, R. Zboril, R. Varma, Cu and Cu-based nanoparticles: synthesis and applications in catalysis. Chem. Rev. **116**(6), 3722–3811 (2016)

25. A. Ali, H. Zafar, M. Zia, I. Ul Haq, A. Phull, J. Ali, A. Hussain, Synthesis, characterization, applications, and challenges of iron oxide nanoparticles. Nanotechnol. Sci. **9**, 49–67 (2016)

26. Y. Zhou, C.-K. Dong, L. Han, J. Yang, X. Du, Top-down preparation of active cobalt oxide catalyst. ACS Catal. **6**(10), 6699–6703 (2016)

27. Z. Wang, X. Pan, Y. He, Y. Hu, H. Gu, Y. Wang, Z. Wang, X. Pan, Y. He, Y. Hu, H. Gu, Y. Wang, Piezoelectric nanowires in energy harvesting applications. Adv. Mater. Sci. Eng. **2015**, 1–21 (2015)

28. M. Kot, T. Major, J. Lackner, K. Chronowska-Przywara, M. Janusz, W. Rakowski, Mechanical and tribological properties of carbon-based graded coatings. J. Nanomater. **2016**, 1–14 (2016)

29. European Union Observatory for Nanomaterials, Regulation [Online]. Available: https://euon. echa.europa.eu/regulation. Accessed 22 Sept 2023
30. European Chemicals Agency, Nanomaterials [Online]. Available: https://echa.europa.eu/reg ulations/nanomaterials. Accessed 22 Sept 2023

Chapter 5
Nanoproducts in Building Materials

Paula Porras-Pereira, Juana Esperanza Llorente-García,
and Costas A. Charitidis

Abstract The second chapter is dedicated to an exhaustive elucidation and evaluation of the deployment of nanoproducts in materials integral to the construction sector. This includes a precise definition of the primary nanoproducts prevalent in this sector, accompanied by an in-depth analysis of their most prevalent characterization techniques. Furthermore, the chapter scrutinizes the environmental and health risks entailed in the application of these nanoproducts within the construction domain.

1 Main Nanoproducts in Construction Sector

A variety of MNMs can have beneficial applications in construction that encompass superior structural properties, functional paints and coatings, and high-resolution sensing/actuating devices [1].

As mentioned above, the nanomaterials used in the construction sector are called manufactured nanomaterials, which are those that have undergone manufacturing processes. Following, nanoparticles containing materials used in the construction industry have been identified, defining its properties and applications in the sector (Table 1).

P. Porras-Pereira (✉)
Department of Building Construction I, Higher Technical School of Architecture, University of Seville, Seville, Spain
e-mail: pporras@us.es

J. E. Llorente-García
Department of Sustainable Construction and Architectural Heritage, Technological Centre for Marble, Stone and Materials, Murcia, Spain
e-mail: juana.llorente@ctmarmol.es

C. A. Charitidis
Department of Materials Science and Engineering, School of Chemical Engineering Iroon Polytechniou, National Technical University of Athens (NTUA), Athens, Greece
e-mail: charitidis@chemeng.ntua.gr

© The Author(s) 2025

P. Mercader-Moyano and P. Porras-Pereira (eds.), *Life Cycle Analysis Based on Nanoparticles Applied to the Construction Industry*,
https://doi.org/10.1007/978-3-031-79115-4_5

Table 1 Nanomaterials in construction [1–6]

Nanoparticle type	Material/ Application	Expected benefits
SiO_2 nanoparticles [1, 2]	Concrete Ceramic Coatings Painting Windows	Reinforcement in mechanical strength; rapid hydration; durability. Coolant; light transmission; fire resistant. Refrigerant properties; anti-reflective; fire resistant. Improves adhesion and durability. Flame-proofing; anti-reflection
TiO_2 nanoparticles [1, 2]	Cement Asphalt Wood Windows Solar Cell Coating/ painting	Rapid hydration; increased degree of hydration; self-cleaning. Durability; noise reduction; reduction of pollutants in the air. Ultraviolet (UV) radiation protection. Superhydrophilicity; anti-fogging; fouling-resistance. Non-utility electricity generation Photocatalytic; self-cleaning
Carbon nanotubes [1, 2]	Concrete Ceramic NEMS/ MEMS Solar Cell	Mechanical durability, crack prevention. Enhanced mechanical and thermal properties Real-time structural health monitoring Effective electron mediation
Fe_2O_3 nanoparticles [1, 2]	Concrete	Increased compressive strength, abrasion-resistant
Cu nanoparticles [1, 2, 5]	Steel	Weldability, corrosion resistance, formability
Ag nanoparticles [1]	Coating/ painting Windows	Biocidal activity. Self-cleaning
Clay nanoparticles [2]	Bricks Mortars	Increased compressive strength and surface roughness
Zycosoil® [2]	Asphalt Concrete	Increased fatigue life, higher compaction
Al_2O_3 nanoparticles [2]	Asphalt Concrete Timber	Increased serviceability
ZnO nanoparticles [4, 5]	Cement Coating/ painting	Enhanced performance. Photocatalytic; self-cleaning; durability; biocidal; hydrophobic
$CaCO_3$ nanoparticles [5]	Concrete	Accelerated hydration, increased the flowability & increased compressive strength
MgO nanoparticles [6]	Coating/ painting	Energy-saving

- *SiO_2*

Silicon oxide (SiO_2), or also known as silica, is nature's most abundant mineral and is found in sand. In their nano form, nanosilica (NS), they are defined as nanoparticles (1–5001 nm) of amorphous SiO_2 insoluble in water. Thanks to its properties, nano silica has become the most reactive silica material, being able to obtain benefits such as increased mechanical strength, flexibility, scratch and abrasion resistance. Size, size distribution and specific surface area are parameters that are defined according to the synthesis process [1, 2].

- *TiO_2*

Titanium oxide or TiO_2 is a natural mineral very abundant on planet Earth, whose physical appearance consists of a white crystalline powder and which is extracted from ilmenite, rutili and anatase. TiO_2 nanoparticles are composed of titanium (Ti) and oxygen (O) atoms, they can come in different shapes and sizes, including spherical nanoparticles, nanotubes and nanofibers. Together with nano-SiO_2, they are the most commonly used additives in building materials. What most characterizes nano-TiO_2 is for its great effectiveness in self-cleaning concrete and providing the additional benefit of decontaminating the environment [1, 2].

- *Carbon nanotubes*

Nanotubes (CNTs) have proven their value in building materials and motivation to find an effective solution to harness their long-term efficiency. CNTs are classified as one-dimensional materials, are allotropes of carbon, cylindrical nanostructure in shape, and are members of the fullerene structural family. It is an element that is composed solely of carbon atoms, with elongated formations with a diameter of 1–100 nm and a length of 100 μm. Nanotubes are classified as single-walled nanotubes (SWNT) and multi-walled nanotubes (MWNT). Carbon nanotubes are perhaps one of the most promising nanomaterials, commonly used as a selective sorbent to remove organic matter and/or biological contaminants from streams. They are allotropes of carbon, cylindrical in shape, have high thermal, electrical conductivity and mechanical properties. In terms of improving flexibility and increasing strength, carbon nanotubes are the best nanomaterials. The main techniques used for production is chemical vapor deposition. This method uses the catalytic transformation of carbon [1, 2].

- *Fe_2O_3*

Iron oxide (Fe_2O_3) is a compound in the form of dust that does not conduct electricity and whose colour depends on the change of electrons at the penultimate energy level. At the nanometre scale, it is a material widely used in the construction industry. Its properties consist of high chemical stability, corrosion resistance and magnetic properties, in addition to exhibiting photocatalytic properties, which means that it can break down organic pollutants under sunlight or artificial light. It features a reduced particle size, which leads to a larger surface area and higher chemical reactivity [1, 2].

- *Ag*

Silver nanoparticles, typically range in size from 1 to 100 nm, are commonly referred to as "silver". Some of these nanoparticles consist primarily of silver oxide, owing to the significant presence of silver atoms on both their surface and within their volume. Various forms of nanoparticles can be synthesized based on specific applications; while spherical shapes are prevalent among silver nanoparticles, thin, octagonal, and diamond-shaped sheets are also frequently encountered [1].

- *AL_2O_3*

These are nano-sized spherical particles of an alumina mass. Alumina nanoparticles can be synthesized through various methods, such as pyrolysis, sputtering, sol-gel, and the commonly preferred technique of laser ablation. Aluminium nanoparticles typically exist in two forms: individual particles with nearly spherical shapes or as oriented fibres [2].

The properties of bulk alumina that render it a versatile element in industry differ from those of its nanoparticles. The larger diameter of bulk aluminium limits its surface interaction with the environment, thereby constraining its potential applications in areas such as catalysis or adsorption. However, the emergence of nanoparticles presents a new frontier for exploring the diverse applications of the already abundant and useful aluminium oxide compound. The reduction in size offers several advantages, but it is the alteration in the chemistry of nanoparticles compared to their larger counterparts that sparks interest in incorporating these nanoparticles across various fields of science and engineering [2].

- *ZnO*

Zinc oxide is a semiconductor material found in nature in the mineral known as Zincite. It stands out as one of the most extensively studied semiconductors in contemporary research, primarily owing to its exceptional physical properties, obtaining its greatest technological significance in low-dimensional structures, such as nanoparticles, nanowires, or nanofibers [4, 5].

In its nanoparticle form, zinc oxide exhibits several key characteristics, such as excellent light retention, low resistivity, non-toxicity, and natural abundance. These properties are particularly important for optoelectronic and piezoelectric materials, where ZnO nanoparticles can be utilized to enhance performance and efficiency [4, 5].

2 Nanomaterial Characterization Techniques

Nanomaterials are characterized using a variety of techniques to understand their properties, structure, composition, and behaviour at the nanoscale.

Nanostructures have garnered significant attention as a rapidly expanding category of materials with diverse applications. Various techniques have been employed to characterize nanoparticles, encompassing aspects such as size, crystal structure, elemental composition, and a range of other physical properties. In many instances, multiple techniques can be utilized to assess the same physical properties. However, the differing strengths and limitations of each technique complicate the selection of the most suitable method, while often a combinatorial characterization approach is necessary to comprehensively evaluate nanoparticles [7].

In recent years, there has been a notable increase in the synthesis of various types of nanomaterials, produced in larger quantities than ever before, thus requiring a pressing need to develop more precise and credible protocols for their characterization. However, achieving comprehensive characterization can be challenging due to inherent difficulties in analysing nanoscale materials compared to bulk materials. Moreover, the multidisciplinary nature of nanoscience and nanotechnology often limits access to a wide range of characterization facilities for research teams. As a result, a comprehensive approach is often required, combining multiple techniques in a complementary manner for a thorough characterization of nanoparticles. It is crucial to understand the strengths and limitations of different techniques to determine if the use of one or two methods alone can provide reliable information when studying a specific parameter. The scientific community recognizes that analytical characterization methods for nanomaterials may operate differently compared to their traditional use for macroscopic materials, highlighting the ongoing growth and evolution of nanoscience and nanotechnology [7, 8].

In this context, various methods for characterizing nanoparticles are thoroughly examined. Some of these techniques are tailored specifically for studying certain properties, while others are versatile and can be combined for a comprehensive analysis [9]. Following, these techniques were compared, taking into account factors such as availability, cost, selectivity, precision, non-destructive nature, simplicity, and affinity to particular compositions or materials [7].

Microscopy-based methods offer insights into the size, morphology, and crystal structure of nanomaterials. Additionally, specialized techniques cater to specific material groups, such as magnetic techniques. Furthermore, many other methods furnish detailed information regarding the structure, elemental composition, optical properties, and various other common and specialized physical properties of nanoparticle samples [7].

Tables 2 and 3 present numerous distinct characterization techniques for NPs, categorized according to the properties studied by each method.

Size and shape represent two of the main parameters examined during the characterization of nanoparticles. Additionally, measurements encompass size distribution, degree of aggregation, surface charge, surface area, and, to some extent, assessment of surface chemistry. It is noteworthy that the size, size distribution, and presence of organic ligands on the particle surface can significantly influence other properties and potential applications of the NPs [7].

However, the analysis of nanomaterials has significant challenges due to the interdisciplinary nature of the field, the lack of suitable reference materials for calibrating analytical tools, complexities associated with sample preparation, and the interpretation of data [7].

- *X-ray-based techniques*

X-ray diffraction (XRD) stands out as one of the most widely employed techniques for characterizing NPs. Generally, XRD furnishes valuable insights into the crystalline structure, phase nature, lattice parameters, and crystalline grain size of the materials under investigation [7].

Table 2 Summary of the experimental techniques that are used for nanoparticle characterization [7]

Technique	Main information derived
XRD (group: X-ray based techniques)	Crystal structure, composition, crystalline grain size
XAS (EXAFS, XANES)	X-ray absorption coefficient (element-specific)—chemical state of species, interatomic distances, Debye–Waller factors, also for non-crystalline NPs
SAXS	Particle size, size distribution, growth kinetics
XPS	Electronic structure, elemental composition, oxidation states, ligand binding (surface-sensitive)
FTIR (group: further techniques for structure / composition / main properties)	Surface composition, ligand binding
NMR (all types)	Ligand density and arrangement, electronic core structure, atomic composition, influence of ligands on NP shape, NP size
BET	Surface area
TGA	Mass and composition of stabilizers
LEIS	Thickness and chemical composition of self-assembled monolayers of NPs
UV-Vis	Optical properties, size, concentration, agglomeration state, hints on NP shape
PL spectroscopy	Optical properties—relation to structure features such as defects, size, composition
DLS	Hydrodynamic size, detection of agglomerates
NTA	NP size and size distribution
DCS	NP size and size distribution
ICP-MS	Elemental composition, size, size distribution, NP concentration
SIMS, ToF-SIMS, MALDI	Chemical information (surface-sensitive) on functional group, molecular orientation and conformation, surface topography, MALDI for NP size
SQUID-nanoSQUID (group: magnetic nanomaterials)	Magnetization saturation, magnetization remanence, blocking temperature
VSM	Similar to SQUID through M–H plots and ZFC-FC curves
Mössbauer	Oxidation state, symmetry, surface spins, magnetic ordering of Fe atoms, magnetic anisotropy energy, thermal unblocking, distinguish between iron oxides

(continued)

Table 2 (continued)

Technique	Main information derived
FMR	NP size, size distribution, shape, crystallographic imperfection, surface composition, M values, magnetic anisotropic constant, demagnetization field
XMCD	Site symmetry and magnetic moments of transition metal ions in ferro- and ferri-magnetic materials, element specific
Superparamagnetic relaxometry	Core properties, hydrodynamic size distribution, detect and localize superparamagnetic NPs
TEM (group: microscopy techniques)	NP size, size monodispersity, shape, aggregation state, detect and localize/quantify NPs in matrices, study growth kinetics
HRTEM	All information by conventional TEM but also on the crystal structure of single particles. Distinguish monocrystalline, polycrystalline and amorphous NPs. Study defects
Liquid TEM	Depict NP growth in real time, study growth mechanism, single particle motion, superlattice formation
Cryo-TEM	Study complex growth mechanisms, aggregation pathways, good for molecular biology and colloid chemistry to avoid the presence of artefacts or destroyed samples
Electron diffraction	Crystal structure, lattice parameters, study order–disorder transformation, long-range order parameters
STEM	Combined with HAADF, EDX for morphology study, crystal structure, elemental composition. Study the atomic structure of hetero-interfaces
Aberration-corrected (STEM, TEM)	Atomic structure of NP clusters, especially bimetallic ones, as a function of composition, alloy homogeneity, phase segregation
EELS (EELS-STEM)	Type and quantity of atoms present, chemical state of atoms, collective interactions of atoms with neighbours, bulk plasmon resonance
Electron tomography	Realistic 3D particle visualization, snapshots, video, quantitative information down to the atomic scale
SEM-HRSEM, T-SEM-EDX	Morphology, dispersion of NPs in cells and other matrices/supports, precision in lateral dimensions of NPs, quick examination–elemental composition

(continued)

Table 2 (continued)

Technique	Main information derived
EBSD	Structure, crystal orientation and phase of materials in SEM. Examine microstructures, reveal texture, defects, grain morphology, deformation
AFM	NP size and shape in 3D mode, evaluate degree of covering of a surface with NP morphology, dispersion of NPs in cells and other matrices/supports, precision in lateral dimensions of NPs, quick examination–elemental composition
MFM	Standard AFM imaging together with the information of magnetic moments of single NPs. Study magnetic NPs in the interior of cells. Discriminate from non-magnetic NPs

XAS quantifies the X-ray absorption coefficient of a material relative to energy. Each element exhibits a distinct set of absorption edges corresponding to various binding energies of its electrons, conferring element selectivity to XAS. Notably, EXAFS is particularly sensitive, offering a convenient means to discern the chemical state of species, even at very low concentrations [7, 10].

The small-angle X-ray scattering (SAXS) technique is utilized to ascertain particle size, size distribution, and shape. However, it is important to acknowledge that SAXS is a low-resolution technique. In certain instances, additional studies using XRD and/or electron diffraction techniques are essential for comprehensive characterization of NPs [7].

- *Additional techniques for the characterization of the structure, composition, and other main NP properties*

Additionally, numerous other techniques help in determining the structure, composition, size, and other fundamental characteristics of NPs [7].

Fourier transform infrared spectroscopy (FTIR) relies on measuring the absorption of electromagnetic radiation within the mid-infrared region (4000–400 cm^{-1}). When a molecule absorbs IR radiation, its dipole moment undergoes modification, rendering the molecule IR active. The recorded spectrum reveals bands corresponding to the strength and nature of bonds, as well as specific functional groups. This information sheds light on molecular structures and interactions [11].

Nuclear magnetic resonance (NMR) spectroscopy is a crucial analytical technique for both quantitative and structural determination of nanoscale materials. It is based on the NMR phenomenon displayed by nuclei possessing non-zero spin when subjected to a strong magnetic field, resulting in a small energy difference between the 'spin-up' and 'spin-down' states. Transitions between these states are probed using electromagnetic radiation in the radio wave range. NMR is commonly employed to investigate interactions or coordination between ligands and the surface of diamagnetic or antiferromagnetic NPs. However, it is unsuitable for characterizing

Table 3 Parameters needed to be determined and the corresponding characterization techniques [7]

Entity characterized	Characterization techniques suitable
Size (structural properties)	TEM, XRD, DLS, NTA, SAXS, HRTEM, SEM, AFM, EXAFS, FMR, DCS, ICP-MS, UV-Vis, MALDI, NMR, TRPS, EPLS, magnetic susceptibility
Shape	TEM, HRTEM, AFM, EPLS, FMR, 3D-tomography
Elemental-chemical composition	XRD, XPS, ICP-MS, ICP-OES, SEM-EDX, NMR, MFM, LEIS
Crystal structure	XRD, EXAFS, HRTEM, electron diffraction, STEM
Size distribution	DCS, DLS, SAXS, NTA, ICP-MS, FMR, superparamagnetic relaxometry, DTA, TRPS, SEM
Chemical state–oxidation state	XAS, EELS, XPS, Mössbauer
Growth kinetics	SAXS, NMR, TEM, cryo-TEM, liquid-TEM
Ligand binding / composition / density / arrangement / mass, surface composition	XPS, FTIR, NMR, SIMS, FMR, TGA, SANS
Surface area, specific surface area	BET, liquid NMR
Surface charge	Zeta potential, EPM
Concentration	ICP-MS, UV-Vis, RMM-MEMS, PTA, DCS, TRPS
Agglomeration state	Zeta potential, DLS, DCS, UV-Vis, SEM, Cryo-TEM, TEM
Density	DCS, RMM-MEMS
Single particle properties	Sp-ICP-MS, MFM, HRTEM, liquid TEM
3D visualization	3D-tomography, AFM, SEM
Dispersion of NP in matrices/supports	SEM, AFM, TEM
Structural defects	HRTEM, EBSD
Detection of NPs	TEM, SEM, STEM, EBSD, magnetic susceptibility
Optical properties	UV-Vis-NIR, PL, EELS-STEM
Magnetic properties	SQUID, VSM, Mössbauer, MFM, FMR, XMCD, magnetic susceptibility

ferri—or ferromagnetic materials due to the large saturation magnetization, which induces variations in local magnetic fields, resulting in frequency shifts and dramatic reductions in relaxation times [12].

The Brunauer–Emmett–Teller (BET) technique is employed for characterizing nanoscale materials, relying on the physical adsorption of a gas onto a solid surface. This method is extensively utilized for determining the surface area of nanostructures, as it offers a relatively accurate, rapid, and straightforward approach for this purpose [12].

Thermal gravimetric analysis (TGA) offers insights into the mass and composition of stabilizers. In this technique, a nanomaterial sample undergoes heating, causing components with varying degradation temperatures to decompose and vaporize, leading to a recorded change in mass. The TGA device records both temperature and mass loss, enabling the determination of the type and quantity of NP organic ligands, considering the starting sample mass. TGA is advantageous for its simplicity and directness, requiring minimal sample preparation aside from ensuring the sample is in a dry state [7].

Nowadays, a multitude of alternative techniques are available for nanoparticle characterization. Among these, the Low-energy ion scattering (LEIS) method stands out as a modern analytical approach facilitating rapid thickness characterization of self-assembled monolayers. Additionally, the UV-Vis spectroscopy (UV-Vis) technique is employed to gauge the intensity of light reflected from a sample, comparing it with a reference material. Photoluminescence (PL) spectroscopy monitors light emitted from atoms or molecules upon photon absorption, particularly useful for characterizing fluorescent nanoparticles. Lastly, the Dynamic light scattering (DLS) technique is widely utilized for sizing nanoparticles in colloidal suspensions within the nano- and submicrometer ranges [7].

- *Characterization methods for magnetic nanostructures*

In this section the focus is on the characterization techniques that are employed to evaluate the magnetic properties of Magnetic NPs.

One such technique is Superconducting Quantum Interference Device (SQUID) magnetometry, which serves as a tool for measuring the magnetic properties of nanoscale materials. Nanomaterials, in particular, exhibit distinct properties compared to bulk materials due to their small size and sensitivity to local conditions. Typical SQUID measurements provide information such as magnetization saturation (MS), magnetization remanence (MR), and blocking temperature (TB). Additionally, SQUID can also be employed to measure the magnetic response of individual molecules [11].

Vibrating Sample Magnetometry (VSM) is an alternative method used to record M-H loops for magnetic nanomaterials, enabling the determination of parameters such as MS and MR. This technique allows for the study of magnetic properties of nanoparticles as a function of magnetic field, temperature, and time [7].

Mössbauer spectroscopy is an invaluable analytical technique grounded in the recoil-free resonance fluorescence of γ-photons within matter containing Mössbauer-active elements, like Fe. This method allows for the assessment of various properties of Fe atoms in nanoparticle samples, including oxidation state, symmetry, spin state, and magnetic ordering. Consequently, Mössbauer spectroscopy aids in identifying magnetic phases within a sample. Additionally, for magnetically ordered materials, Mössbauer spectra recorded across temperature variations enable the estimation of magnetic anisotropy energy and the quantification of thermal unblocking [7].

Ferromagnetic Resonance (FMR) is a spectroscopic technique utilized to investigate the magnetization of ferromagnetic materials, including those at the nanoscale. It shares similarities with Electron Paramagnetic Resonance (EPR) and Nuclear Magnetic Resonance (NMR), as it probes the sample magnetization arising from the magnetic moments of dipolar-coupled yet unpaired electrons. FMR spectra offer valuable insights into the average shape and size of catalyst particles, comprised of ferromagnetic elements, utilized in the production of carbon nanotubes [13].

X-ray Magnetic Circular Dichroism (XMCD) serves as a technique utilized to locally probe and study the site symmetry and magnetic moments of transition metal ions within ferro- and ferrimagnetic materials [7].

- *Microscopy techniques for NP characterization*

Transmission Electron Microscopy (TEM) is a microscopy technique leveraging the interaction between a uniform current density electron beam and a thin sample. Upon reaching the sample, some electrons are transmitted while others are elastically or inelastically scattered. The extent of interaction is influenced by factors such as size, sample density, and elemental composition. The resulting image is constructed from the information gathered from the transmitted electrons [14, 15].

High-Resolution Transmission Electron Microscopy (HRTEM) represents an imaging mode of transmission electron microscopy employing phase-contrast imaging, where both transmitted and scattered electrons are merged to generate the image [16]. Given its capability to furnish crucial information regarding nanoparticle structure, HRTEM has emerged as the predominant technique for characterizing the internal structure of nanoparticles [7].

Electron Diffraction (ED), alternatively referred to as Selected Area Electron Diffraction (SAED), stands as another pivotal microscopy tool for scrutinizing the crystal structure of NPs [17]. This technique is instrumental in probing multiply twinned structures and dynamical events occurring within metal NPs [7].

Electron Energy Loss Spectroscopy (EELS) encompasses a range of techniques that gauge the alteration in kinetic energy of electrons after their interaction with a material [7]. Typically employed to discern the atomic structure and chemical properties of a sample, EELS aids in identifying various aspects such as the type and quantity of atoms present, the chemical state of atoms, and collective interactions of atoms with their neighbours [18].

Electron Tomography (ET), a form of 3D electron microscopy, has been developed to address the limitations of other techniques, which typically provide only two-dimensional (2D) projections of three-dimensional (3D) objects. Beyond offering 3D structural insights, electron tomography enables the analysis of chemical composition in 3D by merging tomography principles with analytical TEM techniques. Consequently, electron tomography is recognized as a pivotal tool for comprehending the relationship between the properties and structure of nanoparticles [7].

Indeed, the aforementioned microscopy techniques represent only a subset of the wide array of characterization techniques available today. For instance, Atomic Force Microscopy (AFM) is a microscopy technique renowned for producing three-dimensional images of surfaces at high magnification. Additionally, Electron

Backscatter Diffraction (EBSD) serves as a microstructural-crystallographic characterization technique utilized in the study of crystalline or polycrystalline materials. EBSD aims to comprehend the structure, crystal orientation, and phase of materials, further enriching the toolbox available for material analysis and research [7].

Definitely, it is imperative to recognize that obtaining a comprehensive understanding of the diverse features associated with a nanomaterial typically necessitates the utilization of multiple techniques. Often, evaluating even a single property thoroughly and comprehensively requires the integration of several analytical methods. By employing a combination of techniques, researchers can achieve a more nuanced and comprehensive characterization of nanomaterials, thereby advancing our understanding of their properties and potential applications [7].

3 Environmental Risks Associated

The integration of MNMs into construction materials offers numerous benefits, such as enhanced mechanical properties and durability. However, these advantages may be counteracted by concerns regarding their potential to act as harmful environmental contaminants if released incidentally or accidentally. This emphasizes the importance of conducting proactive risk assessments and establishing regulatory guidelines to ensure the safe utilization and disposal of products containing MNMs [1].

Despite the numerous reported benefits of MNMs across various sectors, it is important to acknowledge the potential health and environmental risks associated with their development and application, which are not yet fully understood. The production of NMs often involves bottom-up processes, including physical and chemical vapor deposition, as well as liquid-phase synthesis. These methods demand substantial energy and material inputs, leading to pollution in the form of effluents and emissions released into the air, water, and soil [19].

Different nanoforms of a chemical may behave differently in the environment, due to the nanospecific properties of the material. Nanomaterials, because of their small size and unique physicochemical properties, present particular challenges in assessing their environmental fate assessment. Factors such as surface modification, either by intentional functionalisation or by adsorption of environmental species, can significantly influence degradation kinetics, mobility, bioavailability, and bioaccumulation potential. In addition, interactions of nanomaterials with environmental matrices and protein corona formation can alter their behaviour and toxicity.

Furthermore, bioaccumulation of nanomaterials in living organisms is a critical aspect that needs thorough examination. The accumulation of nanoparticles in biological tissues can lead to adverse effects on health and biodiversity. Assessing this phenomenon requires an in-depth understanding of the exposure routes, absorption, distribution, metabolism, and excretion of such particles at the nanoscale level.

An unresolved inquiry revolves around whether nano-enabled construction materials can be engineered to maintain both safety and the advantageous properties that render them useful. Prioritizing the principles of industrial ecology and pollution prevention is essential to avert environmental pollution and the associated impacts caused by MNMs. Certain substances can be re-engineered to yield safer, more environmentally friendly, yet still effective products. Hence, it is crucial to decipher the molecular structures and associated properties that render NMs harmful and identify which receptors may face heightened risks [1].

As novel materials are developed and introduced into various applications, it becomes imperative to comprehend their potential mobility and impacts across air, water, soil, and biota. Prioritizing advanced analytical capabilities is crucial for the detection and characterization of MNMs (whether released from or integrated into construction materials) at environmentally relevant concentrations within the intricate environmental and biological matrices. Additionally, environmentally responsible lifecycle engineering of MNMs in construction must be emphasized. While there is considerable enthusiasm regarding the potential of MNMs to bolster our infrastructure, it is essential to acknowledge reasonable concerns about unintended consequences. This underscores the necessity to support research aimed at safe design, production, use, and disposal practices, alongside initiatives promoting recycling, reuse, and remanufacturing, enhancing the sustainability of both the nanotechnology and construction industries [1]. The following table (Table 4) presents some of the adverse environmental effects of the most frequently used nanomaterials in the construction sector:

Therefore, adopting a comprehensive approach like Life Cycle Assessment (LCA) can offer deeper insights into the potential environmental challenges and guarantee the environmental sustainability of NMs. As aforementioned, LCA adheres to internationally standardized methodologies, following the ISO 14040 series, to evaluate the environmental impacts of a product across its entire life cycle. It accomplishes this by scrutinizing the materials utilized and the energy and emissions released into

Table 4 Environmental implications of nanoparticles emissions

Nanoparticle type	Affected cell/Organ /System
Silver nanoparticle (Ag NP)	Inhibits growth [20, 21]
Titanium dioxide (TiO_2)	Inhibition of the growth of certain species [22, 23]
	Oxidative stress [24]
	Phytotoxicity [25]
Iron oxide (Fe_3O_4)	Oxidative DNA damage [26]
	Malformations [27, 28]
Copper dioxide (CuO)	Accumulation toxicity [29, 30]
	Phytotoxicity [31]
Carbon nanotubes (CNT)	Oxidative stress [32, 33]
	Physical interactions [34, 35]
	Mortality [36]
Magnesium oxide (MgO)	Inhibits growth [37–39]

the environment, originally devised to gauge the environmental footprint of products and their respective processes [40].

Previous research has delved into the LCA of nanomaterials, revealing three principal challenges encountered when modelling them within the LCA framework. The initial challenge involves the insufficient utilization of a suitable functional unit capable of encompassing all the distinctive and additional functionalities inherent in nanomaterials. Subsequently, the second challenge pertains to the dearth of transparent Life Cycle Inventory (LCI) data during the production phase of nanomaterials. In many cases, manufacturers refrain from disclosing information concerning materials and energy inputs due to commercial confidentiality concerns. Lastly, the third challenge revolves around the absence of standardized characterization methods for released nanomaterials. These methods are indispensable within the context of Life Cycle Impact Assessment (LCIA) for comprehensively evaluating the environmental implications of nanomaterials throughout their life cycle [40].

4   Health Risks Associated

In order to advance environmentally conscious nanotechnology within the construction sector, it is imperative to not only address the inadvertent environmental repercussions across all phases, including manufacturing, construction, utilization, demolition, and disposal but also to assess the lifecycle impacts of NMs on the well-being of construction personnel and inhabitants [1].

The incorporation of nanomaterials marks a paradigm shift in enhancing product performance. By leveraging nanoparticles and nanocomposites, the mechanical properties of materials have significantly augmented. However, alongside the enhancement in material quality attributed to these nanocomposites, there arises a substantial compromise in the safety of workers. Nanomaterials pose an imperceptible yet potent threat to the health of labourers [41].

Although they offer notable advantages, a significant portion of workers remain unaware of their interaction with nanomaterials, and the potential adverse effects are not yet fully understood. Multiple studies indicate the existence of documented health hazards associated with various manufactured nanomaterials, which, due to their minute size, have the capability to interact at the cellular level [41].

The inhalation of MNMs during coating, moulding, compounding, and integration processes presents a respiratory health hazard to workers. An evaluation of risk conducted on nano-TiO2 revealed that occupational exposure surpassed the permissible threshold solely during the packaging stage. Furthermore, worker exposure to certain MNMs might occur during the production and processing stages prior to their incorporation into products [1].

Hence, it is imperative to conduct periodic air monitoring throughout all stages of manufacturing across the entire operational zones. Scaling up the production of nanomaterials to quantities suitable for construction applications necessitates substantial adjustments and potential alterations in control measures and contingency plans.

Moreover, the absence of comprehensive material descriptors further impedes the formulation and implementation of handling and safety protocols. Certain nanomaterials may exhibit varying material states throughout their life cycles, influencing the likelihood of occupational exposure [1].

The safety and health protocols concerning nanomaterials in the workplace are confronted with numerous knowledge gaps, primarily stemming from limited data on toxicology, health ramifications, and the effectiveness of ventilation systems and personal protective equipment (PPE). Additionally, the absence of established Occupational Exposure Limits (OELs) and the definition of suitable metrics for assessing exposure to nanomaterials further compound these challenges [42].

The health ramifications associated with emissions from the primary nanoparticles utilized in the construction industry are outlined in Table 5.

Table 5 Health implications of Nanoparticles emissions [1, 43–45]

Nanoparticle type	Affected cell/Organ /System
Silver nanoparticle (Ag NP)	Immune system
	Lungs
	Liver
	Brain
	Carcinogenesis [43]
Titanium dioxide (TiO_2)	Vascular System
	Reproductive Organs
	Fibroblast
	Inflammation in lungs
	DNA damage
	Metabolic changes
	Mitochondrial brain lesions [44]
Zinc oxide nanoparticle (ZnO NP)	Carcinogenesis
	Cell death
	Cell proliferation
Iron oxide (Fe_3O_4)	Oxidative DNA damage
Copper zinc ferrite ($CuZnFe_2O_4$)	DNA damage
	Oxidative DNA damage
Carbon nanotubes (CNT)	DNA damage
	Oxidative stress
Copper dioxide (CuO)	Inflammation
	DNA damage
Silica nanoparticles (SiO_2)	Oxidative DNA damage
	Bronchoalveolar carcinoma
Magnesium oxide (MgO)	DNA damage [45]

Despite the lack of detailed information currently available, preventive measures for handling nanomaterials must be specifically tailored to each work situation, considering the types of nanomaterials used and the available information on exposure to them [46]. A thorough understanding of the type of industrial process involved, the properties of the nanomaterials present, the potential exposures—including the frequency and duration of operations—the specific work procedures and the general characteristics of the working environment is essential for the proper selection of such measures.

If the use and generation of health-hazardous nanomaterials cannot be eliminated or replaced, worker exposure should be minimised through engineering controls at source, organisational measures and, as a last resort, personal protective equipment [47]. To effectively implement preventive measures, it is recommended to follow a well-established control hierarchy:

- Process Modification: changes can be made to work procedures, such as decreasing the amount of nanomaterial used in certain activities or replacing nanomaterials in powder form with other presentations where the nanomaterial is in a liquid medium or embedded in a solid matrix.
- Isolation/Confinement: Operations involving a possible release of nanomaterials in the workplace should be carried out in separate facilities or in areas where handling is carried out from a protected space.
- Engineering Control Measures: These measures are aimed at reducing the emission of the pollutant at the point of origin by creating a physical barrier between the worker and the nanomaterial. These measures include localised extraction systems.

In the workplace context where nanomaterials are present, organisational measures play a crucial role in safeguarding the health of workers. Minimising the number of exposed employees is a fundamental principle, thus reducing the risk potential. Limiting the time of exposure is also an effective strategy to reduce the likelihood of adverse incidents. It is equally important to clearly delimit and signpost work areas, using pictograms to warn about the presence of nanomaterials and the corresponding protective measures. Careful management of the amount of particulate nanomaterial in use at any one time is also essential to maintain a safe working environment.

Continuous training and information to workers on potential risks and preventive measures are essential elements of any safety programme. Such training should be provided on a regular basis to refresh knowledge and practices and to update staff on new developments in the field of nanomaterials.

Maintaining work facilities in a clean and tidy condition is a requirement that not only contributes to operational efficiency, but also to safety. Regular cleaning practices should include the use of wet wipes or hoovers equipped with high-efficiency filters to avoid resuspension of airborne particles. Tools such as compressed air or high-pressure water jets, which can disperse nanomaterials, should be avoided.

Collective protection measures are equally vital and should be seriously considered. Extraction cabinets and filtration collection systems with high efficiency HEPA or ULPA filters, designed to enclose and capture contaminants at the point of origin, are critical in this fight against exposure. The integrity of extraction duct systems, especially for nanomaterials that may be more reactive than their larger-scale counterparts, is crucial to prevent releases and ensure system effectiveness.

Although collective measures should be prioritised, the use of individual protection should not be ruled out. For short-term tasks, full, half or quarter masks with P3 particulate filters can be used. For longer exposures, the use of powered filtering devices with P3 filters is recommended. In addition, the use of protective gloves is recommended, which should comply with current regulations for protection against chemicals and microorganisms, and in the case of disposable gloves, the use of two pairs of overlapping gloves is suggested for greater protection. If the nanomaterial is in powder form, appropriate protective clothing should be worn.

References

1. J. Lee, S. Mahendra, P.J.J. Alvarez, Nanomaterials in the construction industry: a review of their applications and environmental health and safety considerations. ACS Nano **4**(7), 3580–3590 (2010)
2. A. Mohajerani, L. Burnett, J.V. Smith, H. Kurmus, J. Milas, A. Arulrajah, S. Horpibulsuk, A. Abdul Kadir, Nanoparticles in construction materials and other applications, and implications of nanoparticle use. Materials **12**(19), 3052 (2019)
3. H. Boostani, S. Modirrousta, Review of nanocoatings for building application. Procedia Eng. **145**, 1541–1548 (2016)
4. M. Kumar, M. Bansal, R. Garg, An overview of beneficiary aspects of zinc oxide nanoparticles on performance of cement composites. Mater. Today: Proc. **43**(2), 892–898 (2021)
5. A.M. Rashad, Effects of ZnO_2, ZrO_2, Cu_2O_3, CuO, $CaCO_3$, SF, FA, cement and geothermal silica waste nanoparticles on properties of cementitious materials—a short guide for civil engineer. Constr. Build. Mater. **48**, 1120–1133 (2013)
6. H. Tang, S. Li, Y. Zhang, Y. Na, C. Sun, D. Zhao, J. Liu, Z. Zhou, Radiative cooling performance and life-cycle assessment of a scalable MgO paint for building applications. J. Clean. Prod. **380**(2), 135035 (2022)
7. S. Mourdikoudis, R.M. Pallares, N.T.K. Than, Characterization techniques for nanoparticles: comparison and complementarity upon studying nanoparticle properties. Nanoscale **10**, 12871–12934 (2018)
8. Y. Wang, D. Wan, S. Xie, X. Xia, C.Z. Huang, Y. Xia, Synthesis of silver octahedra with controlled sizes and optical properties via seed-mediated growth. ACS Nano **7**(5), 4586–4594 (2013)
9. F. Kim, S. Connor, H. Song, T. Kuykendall, P. Yang, Platonic gold nanocrystals. Angew. Chem. **43**(28), 3673–3677 (2004)
10. A.J. Pugsley, C.L. Bull, A. Sella, G. Sankar, P.F. McMillan, XAS/EXAFS studies of Ge nanoparticles produced by reaction between Mg_2Ge and $GeCl_4$. J. Solid State Chem. **184**(9), 2345–2352 (2011)
11. C. Blanco-Andujar, Sodium carbonate mediated synthesis of iron oxide nanoparticles to improve magnetic hyperthermia efficiency and induce apoptosis (Thesis 2014) [Online]. Available: https://www.researchgate.net/publication/311431656_Sodium_carbonate_mediated_synthesis_of_iron_oxide_nanoparticles_to_improve_magnetic_hyperthermia_efficiency_and_induce_apoptosis. Accessed 24 Sept 2023

12. L. Le Trong, Thesis: Water-dispersible magnetic nanoparticles for biomedical applications: synthesis and characterisation, 2011 [Online]. Available: https://livrepository.liverpool.ac.uk/3062809/. Accessed 24 Sept 2023

13. E.-S.M. Durai, K.A. Abdullin, Ferromagnetic resonance of cobalt nanoparticles used as a catalyst for the carbon nanotubes synthesis. J. Magn. Magn. Mater. **321**(24), L69–L72 (2009)

14. Y.W. Jun, J.W. Seo, J. Cheon, Nanoscaling laws of magnetic nanoparticles and their applicabilities in biomedical sciences. Acc. Chem. Res. **41**(2), 179–189 (2008)

15. Y. Pan, S. Neuss, A. Leifert, M. Fischler, F. Wen, U. Simon, G. Schmid, W. Brandau, W. Jahnen-Dechent, Size-dependent cytotoxicity of gold nanoparticles. Small **3**(11), 1941–1949 (2007)

16. D.B. Williams, C.B. Carter, High-resolution TEM, in *Transmission Electron Microscopy* (Springer, Boston, MA, US, 2009), p. 483–509

17. P.A. Buffat Mater, Electron diffraction and HRTEM studies of multiply-twinned structures and dynamical events in metal nanoparticles: facts and artefacts. Mater Chem Phys. **81**(2–3), 368–375 (2003)

18. B. Schaffer, W. Grogger, G. Kothleitner, F. Hofer, Comparison of EFTEM and STEM EELS plasmon imaging of gold nanoparticles in a monochromated TEM. Ultramicroscopy **110**(8), 1087–1093 (2010)

19. I. Khan, K. Saeed, Nanoparticles: properties, applications and toxicities. Arab. J. Chem. **12**(7), 908–931 (2019)

20. I. Maliszewska, Z. Sadowski, Synthesis and antibacterial activity of silver nanoparticles. J. Phys. Conf. Ser. **146**(012024) (2009)

21. N.R. Panyala, E.M. Peña-Méndez, J. Havel, Silver or silver nanoparticles: a hazardous threat to the environment and human health? J. Appl. Biomed. **6**, 117–129 (2008)

22. J. Hou, L. Wang, C. Wang, S. Zhang, H. Liu, S. Li, X. Wang, Toxicity and mechanisms of action of titanium dioxide nanoparticles in living organisms. J. Environ. Sci. **75**, 40–53 (2019)

23. V.K. Sharma, Aggregation and toxicity of titanium dioxide nanoparticles in aquatic environment—a review. J. Environ. Sci. Health Tox. Hazard. Subst. Environ Eng. **44**(14), 1485–1495 (2009)

24. C. Larue, J. Laurette, N. Herlin-Boime, H. Khodja, B. Fayard, A.M. Flank, F. Brisset, M. Carriere, Accumulation, translocation and impact of TiO_2 nanoparticles in wheat (Triticum aestivum spp.): Influence of diameter and crystal phase. Sci. Total Environ. **431**, 197–208 (2012)

25. U. Song, H. Jun, B. Waldman, J. Roh, Y. Kim, J. Yi, E.J. Lee, Functional analyses of nanoparticle toxicity: a comparative study of the effects of TiO_2 and Ag on tomatoes (Lycopersicon esculentum). Ecotoxicol. Environ. Saf. **93**, 60–67 (2013)

26. N. Dissanayake, K. Current, S. Obare, Mutagenic effects of iron oxide nanoparticles on biological cells. Int. J. Mol. Sci. **16**(10), 23482–23516 (2015)

27. V. Madhubala, T. Kalaivani, A. Kirubha, J.S. Prakash, V. Manigandan, H.K. Dara, Study of structural and magnetic properties of hydro/solvothermally synthesized α-Fe_2O_3 nanoparticles and its toxicity assessment in zebrafish embryos. Appl. Surf. Sci. **494**, 391–400 (2019)

28. X. Zhu, S. Tian, Z. Cai, Toxicity assessment of iron oxide nanoparticles in zebrafish (Danio rerio) early life stages. PLoS ONE **7**(9), e46286 (2012)

29. N. Adam, F. Leroux, D. Knapen, S. Bals, R. Blust, The uptake and elimination of ZnO and CuO nanoparticles in Daphnia magna under chronic exposure scenarios. Water Res. **68**, 249–261 (2015)

30. L. Dai, G.T. Banta, H. Selck, V.E. Forbes, Influence of copper oxide nanoparticle form and shape on toxicity and bioaccumulation in the deposit feeder, Capitella teleta. Mar. Environ. Res. **111**, 99–106 (2015)

31. X. Gao, A. Avellan, S. Laughton, R. Vaidya, S.M. Rodrigues, E.A. Casman, G.V. Lowry, CuO Nanoparticle dissolution and toxicity to wheat (Triticum aestivum) in rhizosphere soil. Environ. Sci. Technol. **52**(5), 2888–2897 (2018)

32. F. Schwab, T. Bucheli, L. Lukhele, A. Magrez, B. Nowack, L. Sigg, K. Knauer, Are carbon nanotube effects on green algae caused by shading and agglomeration? Environ. Sci. Technol. **45**(14), 6136–6144 (2011)

33. K. Kwok, K. Leung, E. Flahaut, J. Cheng, S. Cheng, Chronic toxicity of double-walled carbon nanotubes to three marine organisms: influence of different dispersion methods. Nanomedicine **5**(6), 951–961 (2010)
34. X. Zhu, L. Zhu, Y. Chen, S. Tian, Acute toxicities of six manufactured nanomaterial suspensions to Daphnia magna. Nanopart. Occup. Health **11**, 67–75 (2009)
35. P. Ghafari, C. St-Denis, M. Power, X. Jin, V. Tsou, H. Mandal, N. Bols, X. Shirley, Impact of carbon nanotubes on the ingestion and digestion of bacteria by ciliated protozoa. Nat. Nanotechnol. **3**(6), 347–351 (2008)
36. A. Trompeta, I. Preiss, F. Bem-Ami, Y. Benayahu, C. Charitidis, Toxicity testing of MWCNTs to aquatic organisms. RSC Adv. **9**(63), 36707–36716 (2019)
37. Y. Zhu, Y. Tang, Z. Ruan, Y. Dai, Z. Li, Z. Lin, H. Long, $Mg(OH)_2$ nanoparticles enhance the antibacterial activities of macrophages by activating the reactive oxygen species. J. Biomed. Mater. Res. Part A **109**(1) (2021)
38. M. Fu, J. Liang, S. Wang, C. Geng, W. Zhang, T. Wu, The response of microalgae Chlorella sp. To free and immobilized ZrO_2 and $Mg(OH)_2$ nanoparticles: perspective from the growth characteristics. Environ. Eng. Sci. **37**(6) (2020)
39. A.M. Azzam, M.A. Shenashen, M.M. Selim, A.S. Alamoudi, S.A. El-Safty, Hexagonal $Mg(OH)_2$ nanosheets as antibacterial agent for treating contaminated water sources. ChemistrySelect **2**(35), 11431–11437 (2017)
40. N. Nizam, M. Hanafiah, K. Woon, A content review of life cycle assessment of nanomaterials: current practices, challenges, and future prospects. Nanomaterials **11**(12), 3324 (2021)
41. European Chemicals Agency, in *Nanomaterials* [Online]. Available: https://echa.europa.eu/reg ulations/nanomaterials. Accessed 22 Sept 2023
42. Instituto Nacional de Seguridad e Higiene en el Trabajo (INSHT), in *Seguridad y Salud en el Trabajo con Nanomateriales*, 2015. [Online]. Available: https://www.insst.es/docume nts/94886/96076/sst+nanomateriales/bd21b71f-d5ec-4ee8-8129-a4fa58480968. Accessed 24 Sept 2023
43. C. Marambio-Jones, E. Hoek, A review of the antibacterial effects of silver nanomaterials and potential implications for human health and the environment. J. Nanopart. Res. **12**, 1531–1551 (2010)
44. T. Coccini, S. Grandi, D. Lonati, C. Locatelli, U. De Simone, Comparative cellular toxicity of titanium dioxide nanoparticles on human astrocyte and neuronal cells after acute and prolonged exposure. Neurotoxicology **48**, 77–89 (2015)
45. N.V. Zaitseva, M.A. Zemlyanova, M.S. Stepankov, A.M. Ignatova, Scientific prediction of magnesium oxide nanoparticles toxicity and assessment of its hazard for human health. Ekologiya Cheloveka (Hum. Ecol.) **26**(2), 39–44 (2019)
46. A. Pietroiusti, H. Stockmann-Juvala, F. Lucaroni, K. Savolainen, Nanomaterial exposure, toxicity, and impact on human health. Wiley Interdiscip. Rev. Nanomed. Nanobiotechnol. **10**(5) (2018)
47. W. Nazaroff, Exploring the consequences of climate change for indoor air quality. Environ. Res. Lett. **8**(015022) (2013)

Chapter 6
Nanomaterials in Construction and Demolition Waste (CDW) Management

David Caparrós-Pérez, Georgios Zaverdinos, and Jaime Solís-Guzmán

Abstract This third chapter aims to provide a comprehensive understanding of the ecological footprint and implications associated with the life cycle of nanoproducts in the construction sector. This chapter unfurls a detailed examination of the potential environmental ramifications stemming from the disposal of nanoproducts, defining waste treatment policies associated with nanoproducts, and conducting an exploration of their life cycle dynamics, and an analysis of the pollutant emissions engendered by their application.

The significant expansion of the global construction sector has resulted in a substantial increase in the production of construction and demolition (C&D) waste. This category of waste represents the largest portion of waste generated and demands efficient treatment and utilization to meet sustainability objectives. Within C&D waste, numerous economically valuable materials are present, many of which can be repurposed as construction materials. Ideally, these waste materials are processed or treated in close proximity to demolition sites to ensure a consistent supply of raw materials, such as recycled aggregates, for use in construction projects [1].

Processing and treating C&D waste presents several challenges due to the variability in material properties and the bulky nature of the waste [1].

D. Caparrós-Pérez (✉)
Department of Sustainable Construction and Architectural Heritage, Technological Centre for Marble, Stone and Materials, Cehegín, Spain
e-mail: david.caparros@ctmarmol.es

Present Address:
G. Zaverdinos
Delta Materials Process and Innovation Solutions, Sterea Ellada, Greece
e-mail: gzaverdinos@delta-ms.gr

Present Address:
J. Solís-Guzmán
Department of Building Construction II, Higher Technical School of Building Engineering, University of Seville, Seville, Spain
e-mail: jaimesolis@us.es

P. Mercader-Moyano and P. Porras-Pereira (eds.), *Life Cycle Analysis Based on Nanoparticles Applied to the Construction Industry*,
https://doi.org/10.1007/978-3-031-79115-4_6

1 Waste Treatment Policies for Materials Composed of Nanoparticles

In order to discern the prevailing categories of waste containing nanomaterials, it proves advantageous to conduct a comprehensive evaluation of the collective waste streams. It is imperative to acknowledge that the scarcity of quantitative data pertaining to the production of nanomaterials, their integration into goods, and their pathways into diverse waste streams introduces uncertainty regarding whether certain waste streams predominantly harbour higher quantities of nanomaterials compared to others. Nonetheless, in the absence of alternative methods for identifying predominant waste streams containing nanomaterials, estimating overall waste streams could serve as a useful initial step [1].

The largest portion of C&D waste originates from the construction and demolition of buildings, public infrastructure, road planning and maintenance activities. As previously mentioned, in Europe, nanomaterials find widespread application in various construction components such as cement, insulation materials, and paints. Construction materials often incorporate nanoparticles like titanium dioxide, zinc oxide, silicon dioxide, and aluminium oxide. These nanoparticulate construction materials are commonly found in coatings, glass, concrete, steel, insulation, and composites, reflecting the diverse array of nanomaterials utilized, including silica, titanium, metal oxides, carbon nanotubes, nanoclays, and aluminium, among others. The increasing use of nanomaterials in the construction sector implies their presence in the waste streams generated during construction and demolition processes [1, 2].

The second-largest fraction of waste in the C&D sector typically arises from mining and quarrying activities. However, it's noteworthy that there is a lack of publications specifically addressing the presence of nanomaterials in this particular waste stream [1].

The manufacturing sector constitutes the third most significant contributor to waste generation, encompassing a variety of industries responsible for producing industrial wastewater and residual materials containing nanomaterials. Manufacturing operations commonly result in the generation of substantial quantities of contaminated water, which may carry nanomaterials utilized in the production processes [3, 4]. Additionally, these processes often result in the production of sludge, which may also contain nanomaterials. It's important to note that manufacturing activities span across different industries, each contributing differently to waste streams. While the metal industry constitutes a significant portion of industrial waste (29% of manufacturing waste), there is currently no available data on its contribution to waste streams containing nanomaterials [5].

Another significant portion of waste by mass is municipal solid waste, encompassing waste from households, commerce, small businesses, office buildings, institutions, and selected municipal services. This waste category includes a diverse range of discarded products containing nanomaterials that find their way into waste treatment facilities. According to the OECD 2016 report, municipal solid waste was identified as one of the primary sources of nanomaterials [6]. As the use of nano-materials in products continues to rise, their presence in waste streams is expected to increase as well. Some of the most mentioned nanomaterials found in municipal solid waste include silver, titanium, zinc, and carbon-based nanomaterials [1].

Wastewater represents a substantial source of waste in Europe, with both industrial and household activities contributing to its composition. OECD has highlighted wastewater and sewage sludge as significant sources of nanomaterials [6]. Recent studies investigating the pathways of nanomaterial-containing waste through treatment facilities have underscored the importance of wastewater as a source of nano-materials. Analysis of water samples from wastewater treatment plants has revealed the presence of nanoparticles such as nano cerium oxide and titanium oxide orig-inating from anthropogenic activities. Additionally, nanoparticles rich in Ce–La, Fe–Al, Ti–Zr, and Zn–Cu have been detected in water samples, further indicating the prevalence of nanomaterials in wastewater [7–10]. Moreover, nanomaterials have been increasingly observed in outdoor urban environments, including urban runoffs [1].

Sewage sludge consists of solids separated from water during the wastewater treatment process. Traditionally, it finds application in agriculture as a fertilizer, undergoes treatment in incineration plants, or is disposed of in landfills [5]. Sewage sludge is a byproduct arising from both municipal and industrial wastewater treatment processes. Within municipal sewage sludge, nanomaterials stemming from commonly utilized products such as nano-silver, titanium, zinc, and cerium are detectable [11, 12].

The OECD report delineated seven primary sources of waste that may potentially contain nanomaterials [6]:

- Municipal solid waste.
- End-of-life products.
- Sludge and biosolids from wastewater treatment plants.
- Fly ash and bottom ash from incinerators.
- Landfill leachate.
- Household drainage.
- Commercial and industrial sewage.

The report also classified these sources based on the typical waste management process they undergo. After considering the significant waste streams containing nanomaterials and conducting an extensive literature review on nanomaterials in municipal solid waste, an updated version of the OECD list has been formulated (see Table 1). Additionally, the waste management phase of biological treatment has been incorporated into the classification [1].

Table 1 Potential sources of nanomaterials in waste streams [1]

Waste management phase	Waste source of nanomaterials
Biological treatment	Municipal solid waste Sludge and biosolids from wastewater treatments plants
Recycling	Municipal solid waste End-of-life-products Construction and demolition waste
Incineration	Municipal and industrial (manufacturing) solid waste Sludge and biosolids from wastewater treatments plants
Landfilling	Municipal solid waste Fly ash and bottom ash from incinerators (municipal solid waste and sewage sludge) Sludge and biosolids from wastewater treatments plants Construction and demolition waste
Wastewaster treatment	Household sewage Commercial and industrial sewage Landfill leachate Process water from wet scrubber (municipal solid waste) Urban runoff

2 Waste Management of Materials Composed of Nanoparticles

New products incorporating nanomaterials continue to be introduced to the market for various applications, leading to an increasing volume of these products reaching the end-of-life stage and being disposed of. Nanowaste, whether in solid or liquid form, is presently managed like conventional waste and is thus routed through existing waste management channels. Additionally, nanowaste may find its way into the environment through direct littering. The prevention of littering and illegal disposal typically relies on effective waste management practices and educational initiatives. Nevertheless, there remains a lack of comprehensive understanding regarding the environmental release of nanomaterials from waste streams [13].

The disposal of products containing nanomaterials contributes to the generation of nanowaste, which encompasses a diverse array of waste types due to the broad applications of nanomaterials. Nanowaste originates from various sources, including household, industrial, medical, and research waste streams. Despite its prevalence, there is currently no universally accepted definition of "nanowaste". However, one proposed definition describes it as "separately collected or collectable waste materials that contain engineered nanomaterials". The manner in which nanomaterials are incorporated into waste depends on their initial application. For instance, nanomaterials may be tightly integrated into a solid matrix (e.g., carbon nanotubes in display screens) or exist as free particles or agglomerates dispersed in liquids (such as titanium dioxide in certain sunscreens) [14].

Waste management systems across European countries exhibit variations in organization and technology adoption, encompassing diverse methods such as recycling, incineration, and landfilling. Consequently, the treatment of solid nanowaste will vary depending on the specific waste management infrastructure and practices employed within each locality [15–17].

In regions outside of Europe, landfilling often emerges as the primary method for waste disposal [18]. Additionally, nanomaterials may find their way into water sources and subsequently undergo controlled processing within wastewater treatment facilities. For instance, titanium dioxide nanoparticles found in sunscreen may be washed off during showering, while nanoparticle-laden leachate from solid waste can also be subjected to treatment processes [13].

At present, European Waste Legislation does not include specific regulations for the treatment of nanowaste. Consequently, solid waste containing nanomaterials is managed within existing waste management systems as conventional waste material. However, these conventional systems were not initially designed to address the unique characteristics of nanomaterials. Therefore, the presence of nanowaste raises concerns regarding potential emissions during incineration, leaching from landfills, or during recycling processes, which could result in elevated environmental concentrations of nanomaterials [13].

Throughout various treatment stages, nanomaterials could potentially be liberated from the raw waste. Mechanical processes during recycling, such as crushing, shredding, or grinding, may result in the release of nanomaterials from the waste. Similarly, during incineration, where waste is combusted at high temperatures, nanomaterials may become volatilized. In landfilling, the waste matrix undergoes disintegration, potentially leading to the release of particles. Therefore, the extent and rate of nanomaterial release during waste management will largely depend on the specific processes employed [13].

Nanomaterials within nanowaste could potentially be discharged into the environment through standard waste treatment procedures, consequently infiltrating various environmental domains such as air, water, and soil. For instance, mechanical actions during recycling or combustion processes during incineration may propel nanomaterials into the air. Additionally, waterborne release might transpire during recycling operations or over time in landfills where rainwater is not adequately managed. Furthermore, these liberated nanomaterials could seep into the soil, contributing to environmental dissemination [13]

In general, waste treatment operations incorporate mechanisms designed to capture any substances emitted from the waste, employing filtration systems, gas cleaning systems, or sealed landfills. Nevertheless, the degree of effectiveness of these systems in preventing the release of nanomaterials into the environment remains uncertain across all pathways [13].

Thus, the extent of environmental exposure will be influenced not only by the particular waste treatment procedures and release characteristics, but also by the attributes of the nanowaste (whether solid or liquid) and the types of nanomaterials utilized. While a few case studies have suggested a low to medium potential for environmental exposure, it's important to note that generalizing such conclusions is challenging [13, 14].

3 Management and Recycling of Nanoparticle Composite Materials

Although C&D waste is relatively inert, the main environment concern is associated with its high volume, weight and large worldwide annual production [19]. The management of CDW has been recognised as a major global issue due to the associated environmental impacts and constitutes a priority for most environmental programmes worldwide [20]. In response to this, a range of solutions aimed at increasing recycling routes have been explored in the past decades. This is reflected in the large number of studies on C&D waste in the literature, which are largely focused on the physical and mechanical properties as recycled aggregates for civil engineering applications. Applications of crushed concrete and bricks include production of concrete, mortars or tiles, production of cement [19], as cemented pasted backfill [21] or as road pavement material [22] among many others. Many construction wastes such as concrete, paints or steel contain nano-size range material in their formulation [23] and degradation during prolonged exposure may unlock these nano-particulates. Adverse health impacts on residents and waste workers may occur following resuspension of fine particulates and subsequent inhalation.

Despite the considerable attention given to nanomaterials by researchers, industries, and governments, relatively little focus has been placed on developing recycling and waste management strategies for them. There are only a limited number of studies examining the fate of nanomaterials once nano-products reach the end-of-life stage. Considering the already demonstrated environmental and health risks associated with nanomaterials, it is imperative to implement proper waste management and recycling practices to mitigate their potential toxicities. Following the end-of-life of a nanomaterial product, it may undergo recycling, incineration, or landfilling. However, these particles have the ability to disperse through the air, water, and soil, potentially becoming emerging pollutants if not effectively managed, treated, or recycled [24].

Whilst the physicochemical properties of construction wastes are well documented in the literature, the characteristics of nanoparticles in their formulations are not well known. Ultrafine particulates such as carbon nanotubes, SiO_2, TiO_2, Fe_2O_3, Cu or Ag nanoparticles are increasingly used in multiple applications in construction owing to their ability to improve the physical, chemical, and mechanical performance of concrete, ceramic, steel or paint [25]. Nanomaterials enhance physical and chemical properties of concrete during its use phase, and various types have been tested and recommended by researchers [19]. However, these rarely have been examined embedded in a building material matrix. Despite their unique properties and potential benefits, their release to the environment could bring adverse biological and toxicological effects including cell and DNA damage, inflammatory and immune responses, or ROS-induced oxidative stress among others [25]. Further investigation is required to assess the operational lifespan of constructive solutions incorporating nanomaterials and nanoparticles, as well as their end-of-life processes. It's important to evaluate these impacts, especially given that landfills often serve as the final disposal site. It is essential to be aware the toxicity-related characteristics of nanomaterials in construction, such as size, shape, chemical composition, surface properties, agglomeration/aggregation state, and biodegradability. Hazard assessment for each nanomaterial is crucial; for instance, the higher toxicity of nano-ZnO is linked to its dissolution into toxic Zn2 + , unlike insoluble nano-TiO2 and the non-toxic degradation products of nano-SiO2. Similarly, assessing the potential toxicity of nanoparticles in aquatic environments necessitates consideration of critical parameters [26].

The use of nano-TiO2 as coating in buildings has received considerable interests in recent years due to its excellent ability to purify the environment by capturing some of the pollutants in the air and by using its inherent photocatalytic properties to its advantage. The use of TiO2 coating on windowpanes carries a positive effect on acidification potential, eutrophication potential, criteria air pollutants and smog formation potential, while it increases environmental loads in global warming, fossil fuel depletion, water intake, human health, and ecological toxicity. Taking into account the overall environmental performance, findings indicate that the total environmental score of coated glass pane is 0.44, whereas a value for uncoated glass is 0.56. Therefore, the lower the score, the less negative effect the product has on the environment. Consequently, TiO2 coating has a positive effect on air purifying and environment [27]. The relative abundance of spherical TiO2 nanoparticles in concrete and tiles is also worth noting. At C&D wastes crystalline forms such as anatase and rutile have been found. Currently, it is estimated that 10–30% of TiO2, 30% of ZnO, 5–10% of CeO2 used in paint and coatings are formulated as ultrafine particles close to the nanometric scale [28]. Even traditional pigments such as TiO2 contain a nanometric fraction, from which about 36% of the particles are < 100 nm [29].

For the case of nano-ZnO, this is also used mainly in paints for the construction sector. Life cycle analysis of self-cleaning coatings of metal panels containing ZnO nanoparticles has shown improved performance over coatings without ZnO nanoparticles, in most impact categories of an LCA analysis, thus prolonging their service-life. Overall, the benefits from weathering resistance gained from usage of ZnO nanoparticles outweigh the environmental drawbacks in the nanoparticle production stage [30]. Although there is a lack of information in literature, regarding the life cycle of nano-ZnO enhanced cement, there are indications that suggest a positive environmental impact from the addition of ZnO nanoparticles in cement mixtures, due to the improved material performance, and consequential reduction of required maintenance resources [31, 32]. The presence of nanoparticles in the cement mixture, though, poses health risks for those involved in the cement paste preparation, therefore novel mixing, dispersion, and hydration procedures are being developed [19] and their impact needs to be assessed. The lack of information about end-of-life management of ZnO nanoparticles, highlights the need for more accurate LCA data for every stage of their life cycle, and the development of technologies and protocols for nanoparticle retrieval and recycling from construction and demolition waste.

Nano-SiO2 has found extensive use in cement structures. Meng et al. [30], conducted an environmental assessment that compared the concrete strength and service life of natural concrete, recycled concrete, and recycled concrete with nano-SiO2. The concrete specimens were subjected to natural and marine environments. In marine environment, untreated recycled concrete has a shorter lifespan compared to normal concrete due to high levels of chloride, resulting corrosion in rebar. However, this gap narrows in natural conditions. The incorporation of nano-SiO2 into recycled concrete significantly prolongs its lifespan, nearly equalling that of normal concrete. Despite the energy-intensive process of producing nano-SiO2, its application has minimal environmental repercussions and enhances recycled concretes' performance, while extending its lifespan especially in marine environments. Thus, incorporating nano-SiO2 into recycled concrete exhibits significantly greater environmental efficiency compared to the control groups when considering concrete strength and service life and it is a possible route of recycling also for the nanomaterial [30]. Moreover, Li et al., investigated that nano-SiO2 may act as the inert fillers that block the CO2 diffusion [31] and could also react with Ca(OH)2 to form C-S-H, which could reduce the local pH that led to decreased CO2 capture, that potentially counteracts emissions from concrete production. Therefore, during an LCA of a concrete structure with nano-SiO2 the CO2 emissions should be quantified, taking into consideration also the CO2 capture that is achieved [32]. Finally, a recent LCA study has proved that decreasing transportation distances and eliminating concrete disposal contribute to sustainability of recycled concrete [30].

Spherical nanoparticles of magnetite (Fe3O4) are also abundant in concrete and tiles from C&D wastes. Magnetite is widely used for the manufacture of steel, and it is a common mineral in coal fly ash used for the cement manufacture. It is acknowledged that the Fe in nano-magnetite can induce the formation of reactive oxygen species, promoting the oxidative stress of cells; it is also thought to play a role in the extent to which hydrogen-free radicals attack DNA, resulting in mutation and malignant transformations [33].

Literature demonstrates significant variability and uncertainty concerning the health implications, reactivity, ecological impacts, and environmental fate and transport of CNTs, which is added to cement for enhanced compressive and flexural strength [34]. Therefore, it is imperative to assess the toxicity of resulting concrete when CNTs are incorporated to produce high-performance cements. This evaluation is essential to understand how the inclusion of CNTs may alter the environmental characteristics of the cement [35]. Studies have assessed theoretical high-performance cements reinforced with CNTs through LCAs to compare their environmental impact and performance with that of traditional cements. It was found that the incorporation of CNTs significantly increases the environmental footprint of cement production. Incorporating inorganic nanotubes has been proposed as a more viable approach for reducing the environmental impact while retaining the enhanced properties [36]. End-of-life considerations for CNTs and CNT-incorporating materials are pivotal in determining their safe reuse, repurposing, or recycling. In addition, the presence of carbon nanotubes in elongated and amorphous forms was reported in concrete samples of C&D waste [37]. Currently, most CNTs in industrial use are found in batteries or nanocomposites, yet research on the toxicological impact of CNTs released during incineration or leaching from landfills remains limited. Therefore, further investigation into the toxicological effects of CNT disposal is imperative [38].

Modelling studies suggest that metallic nanoparticles may ultimately end up in landfills following their incineration [39]. For instance, when modelling the flows of nano-TiO_2, nano-ZnO, nano-Ag, and carbon nanotubes (CNTs) in the recycling system, it was observed that the majority of these materials are directed to landfills and incineration plants as waste, with only a small portion being recycled into new products. For example, in Switzerland, out of 43 tons/year of TiO_2 entering the recycling system, 23 tons/year are directed to waste incineration plants, while 13 tons/year are sent to landfills. A small portion (2–3 tons/year) may be exported abroad or used in the production of cement and concrete aggregates, and approximately 2% ends up in wastewater. However, there is some positive news as several recovery processes have successfully retrieved significant portions of nanomaterials from recyclable waste streams [40]. For instance, zinc nanoparticles have been recovered up to 99% using thermal procedures involving inert gas condensation and vacuum separation from waste. Similarly, purified carbon nanotubes have been recovered up to 70% from the waste of supercapacitor materials through filtration and extraction techniques [41].

Efforts to recover nanomaterials from waste streams are frequently hindered by inefficiencies or the absence of suitable methods, largely due to constraints imposed by societal norms, financial considerations, product design, recycling technologies, and the inherent challenges of separation dictated by thermodynamics [42]. To address these issues, proposals for disposal and recycling standards specific to nanowaste have been put forward. These standards encompass policies aimed at regulating nanotechnology, adhering to the precautionary principle in the development of nanomaterials, establishing protocols to address potential risks to human health, environmental well-being, and biodiversity, as well as delineating safe commercialization pathways and disposal methods [41].

As nanotechnology continues to advance and expand its scope, effective waste management of nanomaterials emerges as a formidable challenge. Addressing this challenge requires substantial investment in research, raising consumer awareness about waste handling practices, developing industrial recycling capabilities, ensuring cost-effectiveness of waste management strategies, formulating appropriate policies, and fostering political commitment to implement these measures [41]. Improved regulations and implementation of measures to safely contain C&D waste would be advisable to minimise exposure risks to the population. Organised and supported informal waste recycling could help decrease the environmental burden associated with the use of natural resources whilst effectively contribute to reduce poverty and, crucially, address the public health challenges posed by uncontrolled dumping of construction waste [37].

4 Pollutant Emissions of Materials Composed of Nanoparticles

Nanoparticles emerge through both natural phenomena, like volcanic eruptions, erosion, and forest fires. Generally, contamination is linked with human activities, concerning industrial processes such as combustion of charcoal and fuels, and occurrences like dust storms [43]. The toxicity of nanoparticles is predominantly determined by their origin, stability, size, and concentration, as well as the route and extent of exposure. Additionally, factors such as size, bulk and surface chemistry, mass, aggregation, and surface area also influence their toxicity [44]. Additionally, crucial parameters impacting toxicity include aspect ratio, modifications, surface coatings, and crystalline structure [41].

The challenge of toxicity poses a significant obstacle for researchers, engineers, and scientists engaged in the development of nanomaterials. Various forms of toxicity associated with nanomaterials encompass cytotoxicity, dermal toxicity, pulmonary toxicity, genotoxicity, carcinogenic toxicity, liver toxicity, cardiovascular toxicity, haemolytic toxicity, and immune toxicity [45]. Mechanisms underlying the toxic effects of nanomaterials include apoptosis, generation of reactive oxygen species, formation of free radicals, granuloma formation, and heightened inflammatory responses [46]. Nanomaterials are administered to humans via routes such as intravenous, intraperitoneal, and oral methods [47].

Toxicity levels vary depending on the type of nanomaterial, as well as their physiochemical characteristics, including morphology, electrical, magnetic, and optical properties, surface charge, size distribution, surface chemistry, oxidation state, composition, crystalline structure, aggregation, concentration, dispersion state, and synthesis techniques, among others [48]. Moreover, the size and shape of nanomaterials influence cellular uptake, mass diffusivity, sedimentation velocity, attachment efficiency, and deposition velocity on biological surfaces [47]. Different cell types may respond differently to nanomaterials of varying sizes. Toxicity tends to escalate with increasing nanomaterial concentration [49]. Additionally, synthesis methods impact nanomaterial toxicity due to impurity residues and inadequate refinement. Nanomaterials pose risks to human health, potentially affecting the cardiovascular and central nervous systems, disrupting various organ functions, inducing neurotoxicity, or eliciting immunotoxicity responses [46].

Studies investigating the toxicity of nanomaterials have revealed that iron nanoparticles, for instance, can accumulate within organisms, leading to various effects such as apoptosis, generation of reactive oxygen species, and production of oxidative stress. The process of nanomaterial assimilation within organisms can occur through ingestion or inhalation, subsequently disseminating to different organs and tissues, thereby inducing toxicological effects on the organism [50].

Although numerous studies have explored the toxicological effects of nanomaterials on animal and plant cells, investigations into the toxicological impacts of magnetic nanoparticles on plants remain limited. Some researchers have examined the transformation of copper oxide (CuO) nanoparticles upon interaction with plant cells, noting that the toxicity level of nanoparticles is influenced by their conversion upon interaction with plant cell constituents [19].

Factors such as the active surface area, particle size and shape, and functional groups attached to the material's surface directly influence nanoparticle toxicity, particularly in biotechnological applications, owing to their potential to modulate protein mechanisms and binding. Binding interactions may give rise to various biologically distressing compounds, subsequently leading to reduced enzymatic activities and protein degradation. Environmental parameters also significantly influence nanomaterial behaviour, with characteristics such as water hardness or seawater composition, as well as organic compound constituents, impacting material aggregation processes [51]. Further research focusing on the biochemical interactions and kinetics of nanoparticles is essential to advance our understanding of nanoparticle accumulation dynamics [41].

References

1. S. Trivedi, K. Snehal, B. Das, S. Barbhuiya, A comprehensive review towards sustainable approaches on the processing and treatment of construction and demolition waste. Constr. Build. Mater. **393**, 132125 (2023)
2. W. Jones, A. Gibb, C. Goodier, P. Bust, M. Song, J. Jin, Nanomaterials in construction—what is being used, and where? Constr. Mater. **172**(2), 49–62 (2019)
3. D. Jassby, T.Y. Cath, H. Buisson, The role of nanotechnology in industrial water treatment. Nat. Nanotechnol.Nanotechnol. **13**, 670–672 (2018)
4. A. Kunhikrishnan, H.K. Shon, N.S. Bolan, I. El Saliby, S. Vigneswaran, Sources, distribution, environmental fate, and ecological effects of nanomaterials in wastewater streams. Crit. Rev. Environ. Sci. Technol. **45**(4), 277–318 (2015)
5. Eurostat, "Waste statistics," January 2023. [Online]. Available: https://ec.europa.eu/eurostat/statistics-explained/index.php?title=Waste_statistics#Waste_treatment. [Accessed 22 September 2023]
6. Organisation for Economic Cooperation and Development (OECD), Nanomaterials in Waste Streams: Current Knowledge on Risks and Impacts, Paris: OECD Publishing, (2016)
7. Y. Zheng, B. Nowack, Size-specific, dynamic, probabilistic material flow analysis of titanium dioxide releases into the environment. Environ. Sci. Technol. **55**(4), 2392–2402 (2021)
8. S. Rajkovic, N.A. Bornhöft, R. van der Weijden, B. Nowack, V. Adam, Dynamic probabilistic material flow analysis of engineered nanomaterials in European waste treatment systems. Waste Manage. **113**, 118–131 (2020)
9. K. Phalyvong, Y. Sivry, H. Pauwels, A. Gélabert, M. Tharaud, G. Wille, M. F. Benedetti, Occurrence and origins of cerium dioxide and titanium dioxide nanoparticles in the loire river (france) by single particle ICP-MS and FEG-SEM imaging, Front. Environ. Sci., vol. 8, (2020)
10. K. Mehrabi, R. Kaegi, D. Günther, A. Gundlach-Graham, Emerging investigator series: automated single-nanoparticle quantification and classification: a holistic study of particles into and out of wastewater treatment plants in Switzerland. Environ. Sci. Nano **8**, 1211–1225 (2021)
11. P. Hennebert, A. Anderson, P. Merdy, Mineral nanoparticles in waste: potential sources, occurrence in some engineered nanomaterials leachates, municipal sewage sludges and municipal landfill sludges. J. Biotechnol. Biomater. **7**, 261 (2017)
12. A. Gogos, J. Wielinski, A. Voegelin, H. Emerich, R. Kaegi, Transformation of cerium dioxide nanoparticles during sewage sludge incineration. Environ. Sci. Nano **6**, 1765–1776 (2019)
13. DaNa, Nanomaterials in waste, [Online]. Available: https://nanopartikel.info/en/basics/crosscutting/nanomaterials-in-waste/. Accessed 23 September 2023

14. A. Boldrin, S. Hansen, A. Baun, N.I. Bloch Hartmann, T. Fruergaard Astrup, Environmental exposure assessment framework for nanoparticles in solid waste. J. Nanopart. Res. **16**, 2394 (2014)
15. L. Heggelund, S.F. Hansen, T.F. Astrup, A. Boldrin, Semi-quantitative analysis of solid waste flows from nano-enabled consumer products in Europe, Denmark and the United Kingdom– abundance, distribution and management. Waste Manag. **1**(56), 584–592 (2016)
16. Austrian Academy of Sciences, Nano trust-dossier no. 040en: Nanowaste Nanomaterial-containing products at the end of their life cycle, (2014). Online. Available: https://epub.oeaw.ac.at/0xc1aa500e_0x003146a3.pdf
17. A. Caballero-Guzman, T. Sun, B. Nowack, Flows of engineered nanomaterials through the recycling process in Switzerland. Waste Manag. **36**, 33–43 (2015)
18. A. Keller, S. McFerran, A. Lazareva, S. Suh, Global life cycle releases of engineered nanomaterials. J. Nanopart. Res. **15**, 1692 (2013)
19. A. Akhtar, A. Sarmah, Construction and demolition waste generation and properties of recycled aggregate concrete: a global perspective. J. Clean. Prod. **186**, 262–281 (2018)
20. J. Gálvez-Martos, D. Styles, H. Schoenberger, B. Zeschmar-Lahl, Construction and demolition waste best management practice in Europe. Resour. Conserv. Recycl. Conserv. Recycl. **136**, 166–178 (2018)
21. Q. Chen, Q. Zhang, C. Qi, A. Fourie, C. Xiao, Recycling phosphogypsum and construction demolition waste for cemented paste backfill and its environmental impact. J. Clean. Prod. **186**, 418–429 (2018)
22. J. Tavira, J. Jiménez, J. Ayuso, A. López-Uceda, E. Ledesma, Recycling screening waste and recycled mixed aggregates from construction and demolition waste in paved bike lanes. J. Clean. Prod. **190**, 211–220 (2018)
23. M. Baalousha, Y. Yang, M. Vance, B. Colman, S. McNeal, J. Xu, J. Blaszczak, M. Steele, E. Bernhardt, M. Hochella, Outdoor urban nanomaterials: the emergence of a new, integrated, and critical field of study. Sci. Total. Environ. **557–558**, 740–753 (2016)
24. M. Marcoux, M. Matias, F. Olivier, G. Keck, Review and prospect of emerging contaminants in waste—Key issues and challenges linked to their presence in waste treatment schemes: general aspects and focus on nanoparticles. Waste Manag. **33**(11), 2147–2156 (2013)
25. J. Lee, S. Mahendra, P.J.J. Alvarez, Nanomaterials in the construction industry: a review of their applications and environmental health and safety considerations. ACS Nano **4**(7), 3580–3590 (2010)
26. M. Ferreira, E. Soldado, G. Borsoi, M. Mendes, I. Flores-Colen, Nanomaterials applied in the construction sector: environmental, human health, and economic indicators. Appl. Sci. **13**, 23 (2023)
27. H. Babaizadeh, M. Hassan, Life cycle assessment of nano-sized titanium dioxide coating on residential windows. Constr. Build. Mater. **40**, 314–321 (2013)
28. F. Piccinno, F. Gottschalk, S. Seeger, B. Nowack, Industrial production quantities and uses of ten engineered nanomaterials in Europe and the world. J. Nanopart. Res. **14**, 1109 (2012)
29. A. Weir, P. Westerhoff, L. Fabricius, K. Hristovski, N. Goetz, Titanium dioxide nanoparticles in food and personal care products. Environ. Sci. Technol. **46**(4), 2242–2250 (2012)
30. T. Meng, H. Yu, Y. Zhao, S. Ruan, Strength/Service life-oriented environmental assessment of normal and recycled concrete with nano-SiO2 in a natural and a marine environment: a pioneer case study in Zhejiang Province. J. Build. Eng. **64**, 105623 (2023)
31. G. Li, W. Hu, H. Cui, J. Zhou, Long-term effectiveness of carbonation resistance of concrete treated with nano-SiO2 modified polymer coatings. Constr. Build. Mater. **201**, 623–630 (2019)
32. M. Zajac, J. Skibsted, P. Durdzinski, F. Bullerjahn, J. Skocek, M. Ben Haha, Kinetics of enforced carbonation of cement paste. Cement Concr. Res. **131**, 106013 (2020)
33. R. Grazuleviciene, R. Nadisauskiene, J. Buinauskiene, T. Grazulevicius, Effects of elevated levels of manganese and iron in drinking water on birth outcomes. Polish J. Environ. Stud. **18**(5), 819–825 (2009)
34. M. Fakharpour, M. Gholizadeh Arashti, Effect of three MWCNTs-COOH (chiral, zigzag and armchair) on the mechanical properties and water absorption of cement-based composites. Fullerenes Nanotubes Carbon Nanostruct. **31**(5), 404–411 (2023)

35. O.V. Kharissova, L. Torres Martínez, B. Kharisov, Recent trends of reinforcement of cement with carbon nanotubes and fibers. Adv Carbon Nanostruct. **5**, 137–160 (2016)
36. A. Matanza, G. Vargas, I. Leon, M. Pousse, N. Salmon and C. Marieta, Life cycle analysis of standard and high-performance cements based on carbon nanotubes composites for construction applications, in World Sustainable Building 2014 Conference, Barcelona, (2014)
37. M. Oliveira, M. Izquierdo, X. Querol, R. Lieberman, B. Saikia, L. Silva, Nanoparticles from construction wastes: a problem to health and the environment. J. Clean. Prod. **219**, 236–243 (2019)
38. M. Kim, D. Goerzen, P. Jena, E. Zeng, M. Pasquali, R. Meidl, D. Heller, Human and environmental safety of carbon nanotubes across their life cycle. Nat. Rev. Mater. **9**(1), 63–81 (2024)
39. N. Mueller, J. Buha, J. Wang, A. Ulrich, B. Nowack, Modeling the flows of engineered nanomaterials during waste handling. Environ. Sci. Process. ImpactsSci. Process. Impact. **15**(1), 251–259 (2013)
40. T. Dutta, K. Kim, A. Deep, J.E. Szulejko, K. Vellingiri, S. Kumar, E.E. Kwon, S. Yun, Recovery of nanomaterials from battery and electronic wastes: a new paradigm of environmental waste management. Renew. Sustain. Energy Rev. **82**(3), 3694–3704 (2018)
41. S.A. Mazari, E. Ali, R. Abro, F.S. Khan, I. Ahmed, M. Ahmed, S. Nizamuddin, T.H. Siddiqui, N. Hossain, N.M. Mubarak, A. Shah, Nanomaterials: applications, waste-handling, environmental toxicities, and future challenges–a review. J. Environ. Chem. Eng. **9**(2), 105028 (2021)
42. B. Reck, T. Graedel, Challenges in metal recycling. Science **337**(6095), 690–695 (2012)
43. V. Kumar, A. Choudhary, P. Kumar, S. Sharma, Nanotechnology: nanomedicine, nanotoxicity and future challenges, nanoscience nanotechnology. Asia **9**(1), 64–78 (2019)
44. K. Soto, A. Carrasco, T. Powell, K. Garza, L. Murr, Comparative in vitro cytotoxicity assessment of some manufacturednanoparticulate materials characterized by transmissionelectron microscopy. J. Nanopart. Res.Nanopart. Res. **7**(2–3), 145–169 (2005)
45. Z. Chen, S. Han, S. Zhou, H. Feng, Y. Liu, G. Jia, Review of health safety aspects of titanium dioxide nanoparticles in food application. NanoImpact **18**, 100224 (2020)
46. A. Sukhanova, S. Bozrova, P. Sokolov, M. Berestovoy, A. Karaulov, I. Nabiev, Dependence of nanoparticle toxicity on their physical and chemical properties. Nano Rev. **13**, 44 (2018)
47. M. Akter, T. Sikder, M. Rahman, A. Ullah, K. Binte Hossain, S. Banik, T. Hosokawa, T. Saito, M. Kurasaki, A systematic review on silver nanoparticles-induced cytotoxicity: physicochemical properties and perspectives. J. Adv. Res. **9**, 1–16 (2018)
48. X. Gao, G. Lowry, Progress towards standardized and validated characterizations for measuring physicochemical properties of manufactured nanomaterials relevant to nano health and safety risks. NanoImpact **9**, 14–30 (2018)
49. C. Carlson, S. Hussein, A. Schrand, L. Braydich-Stolle, K. Hess, R. Jones, J. Schlager, Unique cellular interaction of silver nanoparticles: size-dependent generation of reactive oxygen species. J. Phys. Chem. B **112**(43), 13608–13619 (2008)
50. E. Sidiropoulou, K. Feidantsis, S. Kalogiannis, G. Gallios, G. Kastrinaki, E. Papaioannou, M. Václavíková, M. Kaloyianni, Insights into the toxicity of iron oxides nanoparticles in land snails. Comp. Biochem. Physiol. C: Toxicol. Pharmacol. **206–207**, 1–10 (2018)
51. A. Bekdemir, S. Liao, F. Stellacci, On the effect of ligand shell heterogeneity on nanoparticle/protein binding thermodynamics. Colloids Surf. B **174**, 367–373 (2019)

Part III
Integration of Life Cycle Assessment and Nanomaterial Assessment

Chapter 7
Calculation of Life Cycle Assessment of Construction Products with Nanoparticles

Marco Antonio Sánchez-Burgos, Paula Porras-Pereira, and Pilar Mercader-Moyano

Abstract This chapter meticulously undertakes a comprehensive Life Cycle Assessment of a product infused with nanoparticles. Each section within this chapter aligns with distinct phases of the methodology, encompassing the definition of study objectives and scope, development of an inventory detailing consumption and emissions, selection and quantification of environmental impact categories, and the subsequent interpretation of obtained results. The chapter concludes with the synthesis of findings, coupled with formulated recommendations for enhancing the integration of nanoparticles in the construction sector.

In general, nanoparticles are included in composite elements, and the same nanoparticle can be associated to different elements, to form different composite elements. In addition to this, nanoparticles properties mostly depend on the shape of the particle, so their characteristics are different to the bulk material [1].

The same way, the behaviour of nanoparticles released to the environment, or adsorbed by humans is different to the corresponding bulk material. Due to their size and shape, these particles react with the environment and tissues different to the corresponding bulk substance, and currently these processes have not been adequately addressed by scientific research [1]. Consequently, nowadays it is challenging to define the risk assessment of most nanoparticles [2].

Present Address:
M. A. Sánchez-Burgos · P. Porras-Pereira (✉) · P. Mercader-Moyano
Department of Building Construction I, Higher Technical School of Architecture, University of Seville, Seville, Spain
e-mail: pporras@us.es

M. A. Sánchez-Burgos
e-mail: msanchez2@us.es

P. Mercader-Moyano
e-mail: pmm@us.es

© The Author(s) 2025

P. Mercader-Moyano and P. Porras-Pereira (eds.), *Life Cycle Analysis Based on Nanoparticles Applied to the Construction Industry*,
https://doi.org/10.1007/978-3-031-79115-4_7

Currently there are available sources that help to conduct the risk assessment of nanoparticles, such as EUNON Nano Data, which is an EU-funded project. These databases are an important source of information to assess the volume of the emissions, the characterisation, and the risk assessment of most used nanoparticles. Regarding risk assessment, EUNON Nano Data include data on psycho-chemical characterisation, toxicological and ecotoxicological assessment of commercial nanomaterials in the European market [2].

Nanomaterials are produced used and disposed of; therefore, life cycle approach can be applied [3]. Furthermore, LCA methodology is considered the best approach to evaluate the potential impacts of engineered nanomaterials alongside their complete life cycle [4, 5]. However, as stated above, significant gaps remain regarding the long-term fate of nanoparticles' environmental emissions and their effects on human health and ecosystem quality, making LCA application to nanoparticles challenging [1].

Life cycle assessment has already been applied to a wide range of products including nanoparticles [6, 4]. These studies results have been assessed and general conclusions regarding this question have been established. Consequently, there are available guidelines to apply LCA methodology to nanomaterials. A reliable example is the Guidance for applying Life Cycle Assessment to nanomaterials by REACH-nano Consortium, a guidance document that is aimed at helping manufacturers and downstream users of engineered nanomaterials to perform a complete risk and environmental assessment taking into account all life cycles and considerations of nanomaterials [5]. Finally, it has to be stressed that, currently the potential impacts of the released nanomaterials in the categories of Human health and Environmental Toxicity are not included in LCA methods, therefore uncertainties and data gaps exist [4–6].

1 Definition of Objectives and Scope of the Study

The initial phase of the Life Cycle Assessment (LCA) methodology involves defining the goal, which encompasses specifying the intended application, study objectives, target audience, and any adopted constraints and assumptions.

The scope of the LCA study provides a detailed description of the system under assessment and its associated analytical parameters. This scope must align with the previously defined goal and includes identifying the Life Cycle stages incorporated into the study.

Following the goal and scope definition, the next steps involve establishing the functional unit, reference flow, and system boundaries. The functional unit represents the service rendered by the products under evaluation, with all resource consumption and emissions quantities referenced to the production of this functional unit. When conducting comparative LCAs between different products or processes, comparisons are made based on the amount of product required to deliver the functional unit.

The reference flow denotes the quantity of product necessary to deliver the functional unit, while system boundaries delineate which flows, such as emissions and resource consumption, are considered within the LCA study. These flows must be necessary for providing the functional unit.

In conjunction with defining the system boundaries, cut-off criteria and allocation rules are established. Cut-off criteria determine which flows are deemed significant for assessing the potential impacts of the product, including considerations of resulting secondary raw materials and the end-of-life stage for the product. Allocation rules come into play when there are multiple co-products resulting from the assessed system, determining the allocation of impacts to each co-product.

For a thorough and precise evaluation of nanomaterials through LCA, adherence to the following procedures is advised to guarantee comprehensive and precise assessments [5]:

- Taking into account that releases and emissions of nanoparticles occur mostly during use phase and end-of-product stage, and they depend on how nanoproducts are managed, the associated potential impacts will be associated to each nanoproduct Life Cycle. Therefore, LCA studies should be focused in nanoproducts instead of nanoparticles [5].
- Nanomaterials provide us with improved products that can develop enhanced or new functions. In most cases the use of nanomaterials enables the reduction in energy, raw materials and emissions. In that sense, it is important to define an adequate functional unit that cover all these advantages [5].
- Environmental and toxicological impacts for nanomaterials occur during all Life Cycle stages, therefore it is advised to extend LCA studies to "cradle to grave" scope for these products [5].
- For stages with little information, it is advised to consider different scenarios in order to conduct a sensitivity analysis [5].

2 Development of an Inventory of Consumption and Emissions

The inventory analysis, encompassing the gathering of all input and output flows within the evaluated product system, constitutes the Life Cycle Inventory (LCI). Clear establishment of system boundaries is essential for accurately computing input (consumption) and output (emission) flows.

Typically, for data collection, employing primary data for core processes and secondary data for ancillary processes is recommended. Primary data is derived from modelling or monitoring of the processes, while secondary data is sourced from existing databases like Ecoinvent, Gabi, or ELCD.

Presently, Life Cycle Inventories (LCIs) for nanomaterials encounter the following challenges [5]:

- Data corresponding to nanomaterials are not included in existing LCA databases. Therefore, these gaps need to be completed, specifically for each study. Otherwise, the study results will not represent the complete process [5]. It advised to prioritise data taken from real measures of assessed processes.
- The information regarding nanomaterials production is often confidential. Furthermore, this is a fast-evolving field of technology. As a consequence, available data is scarce, and it is needed to make estimations in some processes. This leads to some degree of uncertainty that has to be assessed [5]. Each nanoparticle synthesis method assessed has to be studied to obtain consumption and emission flows. For estimations, different scenarios must be evaluated, and discussed [5].
- Regarding emissions of nanoparticles, in most cases a critical step is the incorporation of nanoparticles in the form of powder, to the material. In these cases, the exposition of workers to nanoparticles emissions during production has to be included in the LCA study.
- LCA studies for nanomaterials should encompass all life cycle stages, from production to disposal, adhering to the "cradle to grave" approach.
- In general, there is no information about the released quantities of nanoparticles and their fate in the long term, for each stage of their Life Cycle. Furthermore, there is no consensus about how to measure these emissions.

This data must be gathered on a process-specific basis. It is imperative to acquire extensive information regarding the assessed process to compile a comprehensive set of emission data. The emissions of nanoparticles throughout all stages of the process's life cycle must be incorporated into the resulting LCI.

- Uncertainty must be assessed.
- It is advised to gather information regarding emissions using templates.

In detail, the following data should be addressed:

- Production stage:

  - Inputs and outputs during production stage; consumption and emission associated.
  - Emissions of nanoparticles during production stage, exposition of workers.
  - Releases during production stage, compartment of the emission.
  - Transformation of the particle after the emission

- Use stage:

  - Lifespan and services obtained from the product.
  - Inputs and outputs produced during use; maintenance, cleaning, consumption and emission associated to use.
  - Emissions of nanoparticles during production stage, exposition of workers.
  - Releases during production stage, compartment of the emission. Transformation of the particle after the emission
  - Possibility of nanoparticles emissions during use. Environmental compartment of the emission. Transformation of the particle after the emission

- End-of-life stage.

    - Characteristics of nanoproducts wastes generated at the end-of-life stage.
    - Treatment and final disposal of nanoproducts wastes.
    - Recycling: Type of recycling process. Emissions of nanoparticles during recycling. Quantity of nanoparticles in recycled products.
    - Disposed to landfill: Degradation or transformation of nanoproducts. Environmental compartment for the final fate of nanoproducts waste.
    - Incineration: transformation of nanoparticles after incineration. Nanoparticles included in resulting ashes. Environmental compartment for the final fate of nanoproducts included in resulting ashes.

An example of check-list template for LCI data gathering is included (Table 1) [4, 7].

3   Selection and Quantification of Environmental Impact Categories

This phase encompasses four sequential steps:

- Classification: Assignation of each consumption and emission to the relevant impact category.
- Characterisation: Calculation of impact contribution for each emission and consumption, and aggregation of contributions related to each impact.
- Normalisation (this step is optional): Impact scores al multiplied by normalization factors, which relate the obtained impact for each category to the global impact produced in social group (European, national, global). This allows to inform of the relative relevance of the impacts obtained.
- Weighting (this step is optional): This is a rather controversial step. It consists of multiplying obtained impact scores to enable comparison between impact categories. It has to be stressed that according LCA methodology, scores corresponding to different impact categories cannot be aggregated.

Various methods exist for converting emissions and consumptions of substances into impact scores for different environmental categories. Presently, midpoint impact category methods are well-established, offering robust and meaningful results albeit with challenging interpretation. Examples of these established methods include CML and ReCiPe.

Alternatively, endpoint impact category methods are easier to interpret, yet lack scientific consensus, resulting in limited usage. These methods are not widely adopted due to their less-established nature.

Environmental impact category methods employ scientifically developed models that quantify the relationship between material consumption or substance emissions and the generated impacts.

Table 1 Check-list template for LCI data gathering [7]

	Type of information	Data requested
	Process description	*General description of the process (Ex. Synthesis of LFP nanoportides)*
	Productive process	Typology of process/route of production
	Partner/company responsible	
Resulting material/product description. Flow reference:	Material produced	Type of material synthesized. For *ENMS, specify the format of the final product (powder, dilution,...,)*
		Quantity (g)
	Co-products (If any)	Quantity (g), use of co-product
Process description	Phases of the process	
	Duration (hours)	
	Equipment used	
	Process scale	Scale (lab, pilot, industrial,…)
		Production capacity (kg/year)
Inventory of INPUTS	Energy consumption (Electricity)	Source/origin
		Quantity (kWh)
		Function/use (phase, equipment used,…)
	Energy consumption (heat)	Source/origin
		Quantity (MJ)
		Function/use (phase, equipment used,…)
	Water consumption	Source/origin
		Quantity (1)
		Function/use (process water or cleaning water,…}
	Raw Materials (precursors, gases, solvents, others…] Other materials/substances used within the process, including ancillary materials (cleaning,...)	Name/source
		Quantity (g)
		Funetion/use (precursor, solvent,..,)
		Origin: geographical (km), synthesis process,…
		Other information (supplier, % recycled content…,)
	Packaging	Packaging material (type), weight (g),…
		Size and capacity of packaging

(continued)

Table 1 (continued)

	Type of information	Data requested
	Transport processes inputs	Distance (km), type of vehicle
Inventory OUTPUTS	Direct emissions to air (including ENMs emissions)	Name/type of emission
		Quantity (g)
		Process origin
		Treatment/filtration (% of elimination of ENMs in filtration)
	Emissions to water [including ENMs emissions)	Pollutants, % of ENMs, type of effluent
	Wastewater produced	Type of wastewater (characterisation, pollutants)
		Origin process (cleaning,,.)
		Potential content of ENMs (%of ENMs)
		Treatment/Final Destination (% of degradation of ENMs, elimination and release of ENMs)
	Solid waste	Name/Type
		Classification/code
		Content of ENMs
		Origin process
		Quantity [g]
		Treatment/Destination (ENMs degradation and liberation)
	Liquid waste	Name
		Classification/code
		Content of ENMs
		Origin process
		Quantity [g]
		Treatment/Destination [ENMs degradation and liberation)
	Other outputs (scraps, subproducts, co-products,.)	

It's important to note that current impact category calculation methods do not include nanomaterials. Consequently, there are no defined characterization factors to assess nanoparticle emissions, leading to their exclusion from resulting impact scores.

In general, nanoparticle emissions impact Human Toxicity and Environmental Toxicity categories. However, these categories cannot be adequately addressed for nanoparticle emissions using current impact calculation methods.

Given these considerations, recommendations for nanomaterials LCA should include:

- Utilizing impact methods recommended at the European level, such as those outlined in the ILCD handbook.
- Including relevant categories for nanoparticles, such as Human Toxicity and Ecosystems Toxicity.
- Deriving characterization factors for assessed emissions using prospective approaches based on consensus models aligned with the characteristics of releases and fate of the corresponding process.

To achieve this, it is recommended to collaborate with a diverse team of experts specializing in risk assessment from various fields. If calculating characterization factors proves to be impractical, it is advised to employ a precautionary approach. This involves utilizing characterization factors for analogous substances, such as the corresponding bulk material, while carefully considering potential disparities in final fate behaviour.

3.1 Recommended Impact Category Calculation Methods

The Table (Table 2) features the recommended methods outlined in the ILCD Handbook. Notably, impacts such as Water Depletion and Land Transformation are deemed insignificant for nanoparticle studies and thus are advised to be omitted. However, for other methods, it is imperative to incorporate the specific risks associated with nanoparticles in LCA studies. It is important to note that none of these methods encompass flows or characterization factors tailored to evaluate the specific damages caused by nanoparticles. Hence, the assessment of this risk must be conducted separately [7].

The pertinent impact categories concerning nanoparticles are as follows:

- Ecotoxicity for aquatic fresh water.

For Soil Ecotoxicity and Marine Ecosystems Toxicity, there is insufficient consensus, hence they are not considered. The only method recommended by the ILCD for evaluating Ecotoxicity in Freshwater is the midpoint USEtox.

Currently, no ecotoxicity impact calculation method (including USEtox) incorporates characterization factors for nanoparticles. USEtox includes characterization factors for bulk material emissions to air, water, and soil. To assess the effects of nanoparticle emissions, new characterization factors must be established. It's important to note that the USEtox tool calculates fate factors for emissions based on the behaviour of soluble compounds, thus nanoparticles are not adequately represented in

Table 2 ILCD Handbook recommended impact categories [7]

PEF impact categories	ILCD recommended Impact assessment model	Classification of recommended impact method (ILCD)	Significant for LCA of nanoproduets	Relevant for released Nanopartfcles
1. Climate change	Bern model—Global Warming Potentials (GWP) over a 100 year time horizon	I (recommended and satisfactory)	Potentially significant during all life cycle, especially manufacturing	No
2. Ozone depletion	EDIP model based on ODPs of the World Meteorological Organization (WMO)	I (recommended and satisfactory)	Potentially significant during ail life cycle	No
3. Ecotoxicity for aquatic fresh water	U5Etox model [8]	II (recommended but in need of same improvements)/ III (recommended, but to be applied with caution)	Potentially significant during all life cycle, especially end-of-llife	Yes
4. Human toxicity—cancer effects	USEtox model [8]	II (recommended but in need of some improvements)/ III (recommended, but to be applied with caution)	Potentially significant during all life cycle, especially end-of-life	Yes
5. Human toxicity—oon-cancer effects	U5Etox model [8]	II (recommended but in need of some improvements)/ III (recommended, but to be applied with caution)	Potentially significant during all life cycle, especially end-of-life	Yes

(continued)

Table 2 (continued)

PEF impact categories	ILCD recommended Impact assessment model	Classification of recommended impact method (ILCD)	Significant for LCA of nanoproduets	Relevant for released Nanopartfcles
6. Particulate matter/resp. Inorganics	RiskPoli model [9] and Greco et al. 2007	I (recommended and satisfactory)	Potentially significant during all life cycle	Yes
7. Ionising radiation—HH effects	Human health effect model as developed by Drelcer et al. 1995 [10]	II (recommended but in need of some improvements}	Potentially significant during all life cycle	No
8. Photochemical ozone formation	LOTOS-EUROS model as applied in ReCiPe	III (recommended but in need of some improvements)	Potentially significant during all life cycle	No
9. Acidification	Acumulated Exceedance model [11, 12]	II (recommended but in need of some improvements)	Potentially significant during all life cycle	No
10. Eutrophlcation-terrestrial	Acumulated Exceedance model [11, 12]	II (recommended but in need of some improvements)	Potentially significant during all life cycle	No
11. Eutrophlcation—aquatic	ELTTREND model [13] as implemented in ReCiPe	III (recommended but in need of some improvements)	Potentially significant during all life cycle	No
12. Resource depletion-water	Swiss Scarcity mod. [10]	III (recommended, but to be applied with caution)	Not significant	No

(continued)

Table 2 (continued)

PEF impact categories	ILCD recommended Impact assessment model	Classification of recommended impact method (ILCD)	Significant for LCA of nanoproduets	Relevant for released Nanopartfcles
13 Resource depletion-mineral, fossil	CML2002 model [14]	II (recommended but in need of some improvements)	Potentially significant for manufacturing processes	No
14. Land transformation	Soil Organic Matter (50 M) model [15]	III (recommended, but to be applied with caution)	Not significant	No

this model. Considering nanoparticles' diverse characteristics dependent on physical properties like shape and size, each nanoparticle requires specific examination.

- Human Toxicity

The impact on human health resulting from emissions of chemicals released into the three natural compartments (soil, air, and water) depends on the fate of these emissions, human population exposure, and the toxicological effects on the population.

Currently, the USEtox tool is considered the best option, albeit requiring some enhancements. As mentioned earlier, this tool lacks nanoparticle characterization factors. To assess the effects of nanoparticle emissions, new characterization factors need to be developed.

- Particulate Matter

No comments regarding this impact category.

4 Interpretation of Results

This phase involves a comprehensive review of inventory data and impact scores to draw conclusions from the study. It encompasses sensitivity analysis and uncertainty analysis.

Sensitivity analysis entails calculating impact scores while varying key consumption and/or emission parameters to quantify how modifications to the calculation scenario affect the obtained results. Although not mandatory in ISO standards, it is recommended.

Uncertainty analysis involves statistically assessing the quality of data included in the Life Cycle Inventory. While not compulsory in ISO standards, it is recommended. Currently, this assessment is conducted using tools such as Monte Carlo software.

As mentioned earlier, LCA studies on nanomaterials entail a high level of uncertainties; therefore, all utilized data and sources of uncertainty must be meticulously documented.

Obtaining reliable and robust results from nanoparticles LCA is crucial for enhancing existing methods and databases. Taking this into account, the following guidelines are proposed:

- Conduct an external critical review by a panel of experts in LCA and nanotechnology.
- Perform uncertainty analysis to evaluate the robustness of the results.
- Utilize the results from LCA studies to augment inventory data for existing databases and improve impact characterization methods.

In scenarios where no characterization factor exists for the assessed emissions, a predictive scenario specifically tailored for the assessed nanomaterial should be included. This scenario should encompass:

- Quantification of nanoparticle releases throughout the entire life cycle of the product (from cradle to grave).
- Identification of environmental compartments for each release.
- Prediction of the long-term fate of nanoparticles released, including their transformations in the environment.
- Assessment of toxicity for both humans and the environment due to emitted nanoparticles.

All this data should be included in the LCA study as additional information annexed in the interpretation step.

5 Conclusions and Proposals for Improvement of NPs in the Construction Sector

The estimated size of the EU nanomaterials (NMs) market for 2020 stands at 140.9 Kilotons per volume and € 5205 million per value. Over the next five years, this market is projected to experience substantial growth, with a Compound Annual Growth Rate (CAGR) of 13.9% per volume and 18.4% per value. Given this trajectory, it is imperative to assess the impacts associated with the production and utilization of these products [5].

As discussed in previous sections, there exists a notable dearth of information concerning the fate of nanoparticle emissions upon release into the environment, as well as their ramifications for human health and ecosystem integrity. Consequently, the conventional Life Cycle Methodology alone cannot adequately address the study of nanoproducts. Alternative methodologies capable of evaluating the impacts on human health and ecosystems resulting from nanoparticle emissions must be employed.

Moreover, comprehensive information on emissions throughout the entire life cycle of nanoproducts must be provided for each study, as existing databases for product Life Cycle Inventory lack specific nanoparticle data. In light of the current inability to calculate impacts associated with nanoparticle emissions, it is prudent to incorporate information on predicted emissions and conduct risk assessments for the involved nanoparticles in LCA studies.

Fortunately, several institutions have established databases containing valuable information on nanoparticle emissions, environmental fate, and risk assessment. One notable example is the EU-funded Eunon Nano Data, which offers insights into nanoparticles associated with nanoproducts legalized in the European market. This platform provides accessible information necessary for registering nanoproducts in the European market [5].

References

1. I. Khan, K. Saeed, Nanoparticles: properties, applications and toxicities. Arab. J. Chem. **12**(7), 908–931 (2019)
2. European Chemicals Agency, "Study on the Product Lifecycles, Waste Recycling and the Circular Economy for Nanomaterials," November 2021. [Online]. Available: https://euon.echa.europa.eu/documents/2435000/3268576/nano_lifecycles_euon_en.pdf/107f2bd6-8967-5466-8f48-5610d9120bbe?t=1636969415023. Accessed 22 Sept 2023
3. European Chemicals Agency, "Risk assessment of nanomaterials—further considerations," 2017 [Online] Available: https://euon.echa.europa.eu/documents/2435000/3268584/nano_in_brief_en.pdf/295c5f46-0f1e-4ad5-72a5-81c44b45bdd5?t=1531477424187. Accessed 22 Sept 2023
4. B. Salieri, D. Turner, B. Nowack, R. Hischier, Life cycle assessment of manufactured nanomaterials: where are we? NanoImpact **10**, 108–120 (2018)
5. REACHnano Consortium, "Guidance for applying Life Cycle Assessment methodology to nanomaterials," 2015 [Online] Available: https://invassat.gva.es/documents/161660384/162311778/02+Guidance+for+applying+Life+Cycle+Assessment+methodology+to+nanomaterials/f40ee359-a279-4464-b2d0-c35c17d5922f. Accessed 2023 Sept 25
6. N. Nizam, M. Hanafiah, K. Woon, A content review of life cycle assessment of nanomaterials: current practices, challenges, and future prospects. Nanomaterials **11**(12), 3324 (2021)
7. European Commission, Joint Research Centre—Institute for Environment and Sustainability, "International Reference Life Cycle Data System (ILCD) Handbook—General," 2010
8. R. Rosenbaum, T. Bachmann, L. Gold, et al., USEtox—the UNEP-SETAC toxicity model: recommended characterisation factors for human toxicity and freshwater ecotoxicity in life cycle impact assessment. Int. J. Life Cycle Ass **13**, 532–546 (2008)
9. J. V. Spadaro, A. Rabl, Pathway analysis for population-total health impacts of toxic metal emissions. Risk Anal. Off. Publ. Soc. Risk Anal. **24**(5), 1121–1141 (2004)
10. R. Frischknecht, A. Braunschweig, P. Hofstetter, P. Suter, Human health damages due to ionising radiation in life cycle impact assessment. Environ. Impact Ass. Rev. **20**(2), 159–189 (2000)
11. J. Seppälä, M. Posch, M. E. A. Johansson, Country-dependent characterisation factors for acidification and terrestrial eutrophication based on accumulated exceedance as an impact category indicator. Int. J Life Cycle Ass **11**, 403–416 (2006)

12. M. Posch, J. Seppälä, J. P. Hettelingh, M. Johansson, M. Margni, O. Jolliet, The role of atmospheric dispersion models and ecosystem sensitivity in the determination of characterisation factors for acidifying and eutrophying emissions in LCIA. Int. J. Life Cycle Ass. **13**, 477–486 (2008)
13. J. Struijs, A. Beusen, H. Van Jaarsveld, M. A. J. Huijbregts, Aquatic eutrophication (2009)
14. J. Guinee, Handbook on life cycle assessment operational guide to the ISO standards. Int. J. Life Cycle Ass. **7**, 311–313 (2002)
15. L. Milà i Canals, C. Bauer, J. Depestele, A. Dubreuil, R. Knuchel, G. Gaillard, O. Michelsen, R. Müller-Wenk, B. Rydgren, Key elements in a framework for land use impact assessment within LCA. Int. J. Life Cycle Ass. **12**, 5–15 (2007)

Part IV
Application of Life Cycle Assessment Results in Building Assessment

Chapter 8
EPDs for Construction Products with Nanoparticles

Varun Gowda Palahalli Ramesh and Christina Lee

Abstract This initial segment serves as an introduction, elucidating fundamental knowledge needed to understand and conduct an LCA for the development of EPDs for construction products with nanoparticles. This chapter encompasses an overview of the general framework of Environmental Product Declarations in Europe, detailing the main standards governing EPDs. Additionally, it provides guidelines for developing EPDs for construction products containing nanoparticles, outlining the necessary procedures they must undergo.

Environmental Product Declarations (EPDs) are standardized verified documents containing information on the environmental profile of a product. EPDs are Type III environmental declarations that are useful for business-to-business communication, decision making and comparing the environmental profile of products with similar functions. In the context of the construction products, within the European Union, the EPDs are developed by following the standard EN 15804:A2 (2021) [1]. EPDs are created based on the Life Cycle Assessment (LCA) methodology, which evaluates the environmental performance of a product. LCA is conducted by analysing process inputs (e.g., material and energy-based consumables) and outputs (e.g., waste) during the life cycle of a product. For manufacturers, the manufacturing phase of a product is obligatory to cover. This is the phase where the manufacturer has the most influence and is, therefore, in focus for this part. The core reading material for this part is the regulation EN 15804:2012 + A2:2019/AC:2021—Sustainability of construction

Present Address:
V. G. P. Ramesh (✉) · C. Lee
Department of Industrial & Materials Sciences, Chalmers University of Technology, Gothenburg, Sweden
e-mail: varung@chalmers.se

C. Lee
e-mail: leec@chalmers.se

P. Mercader-Moyano and P. Porras-Pereira (eds.), *Life Cycle Analysis Based on Nanoparticles Applied to the Construction Industry*,
https://doi.org/10.1007/978-3-031-79115-4_8

works—Environmental product declarations—Core rules for the product category of construction products [1].

1 General Framework of Environmental Product Declarations

Several standards need to be followed while developing an EPD for a construction product in Europe, and they are shown in Fig. 1. The first step in developing an EPD is to decide on the framework that will be followed. This is determined by the Programme Operator whom the EPD will be published by. A common programme operator in Europe is the International EPD System. Once a programme operator is decided, the structure that needs to be followed can be investigated which usually consists of several standards. For conducting an LCA study on which the EPD is formed, the methodology is described in the earlier units. To recap, there are two standards, ISO 14040 [2] and ISO 14044 [3], that are followed by LCA practitioners while conducting the LCA.

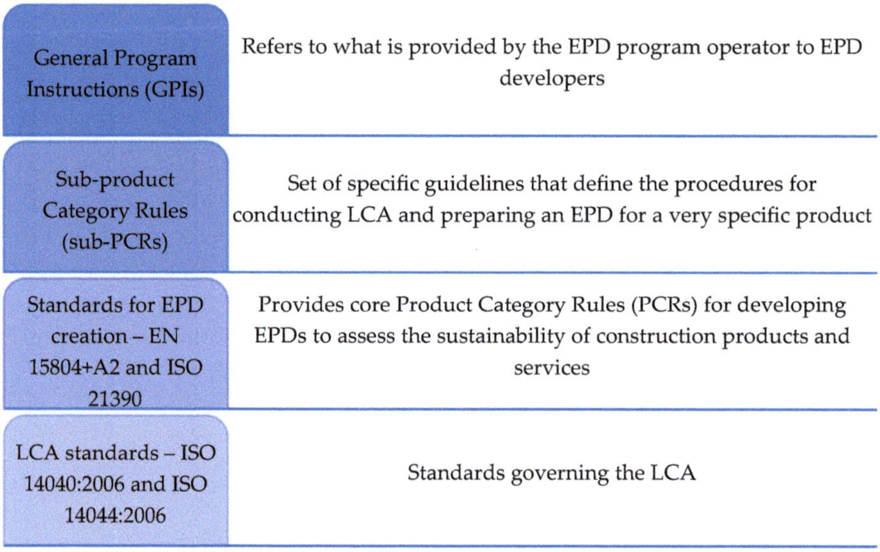

General Program Instructions (GPIs)	Refers to what is provided by the EPD program operator to EPD developers
Sub-product Category Rules (sub-PCRs)	Set of specific guidelines that define the procedures for conducting LCA and preparing an EPD for a very specific product
Standards for EPD creation – EN 15804+A2 and ISO 21390	Provides core Product Category Rules (PCRs) for developing EPDs to assess the sustainability of construction products and services
LCA standards – ISO 14040:2006 and ISO 14044:2006	Standards governing the LCA

Fig. 1 Framework for developing the EPDs for the construction sector in Europe

2 EN 15804:A2 (2021)

In the context of developing EPDs for construction products within the European Union, standard EN 15804 has been developed. The latest version of European standard EN 15804:2012 + A2:2019/AC:2021, referred to as EN 15804:A2 (2021) in this chapter, sets forth guidelines for the allocation of the inputs (such as energy and material) and outputs (waste) to different life cycle modules and for quantifying the environmental performance of construction products [1]. The guidelines are provided in the form of Product Category Rules (PCRs). PCRs provide the rules, requirements, and guidelines to conduct an LCA and develop an EPD for a specific product category. They are used to ensure consistency and comparability among EPDs for functionally similar products. The PCR *2019:14 Construction products within the International EPD System* (based on EN 15804:A2 [4]) provides clear information on how to define the system boundary, choose a functional unit for the assessment, and which impact categories are to be declared.

2.1 Standards Governing EPDs in Detail

The following section provides in-depth information on the standards used for conducting LCAs and EPDs.

ISO 14040 and 14044

In 1993, the International Organization for Standardization (ISO) initiated a standardization process aimed at enhancing the interpretability of LCA results. This initiative resulted in the development of international standards for LCA. Notable among these at that time were ISO 14040, which focused on Life Cycle Impact Assessment (LCIA); ISO 14041, dedicated to Life Cycle Inventory (LCI) modelling; and ISO 14043, aimed at interpreting LCA results. The standardisation process resulted in the development of a common framework and fundamental principles for conducting an LCA study [5]. The standards have since been updated and the following two international standards provide the framework and principles for conducting modern LCA studies:

- ISO 14040 [2]: Environmental management—Life cycle assessment—Principles and framework. The standard describes the "principles and framework for LCA, encompassing: the definition of the goal and scope, the LCI phase, the LCIA phase, and the life cycle interpretation phase".
- ISO 14044 [3]: Environmental management—Life cycle assessment—Requirements and guidelines. The standard specifies "requirements and provides guidelines for LCA, encompassing: defining the goal and scope of the LCA study, the LCI phase, the LCIA phase, the life cycle interpretation phase, reporting and critical review of the LCA results, limitations of the LCA, the connection between LCA phases and the criteria governing the utilization of value choices and optional elements".

GPI

Your chosen EPD programme operator will have a general set of instructions on how an EPD should be produced within their programme, known as the General Programme Instructions (GPI). These build on the ISO standards for describing the LCA methodology with additional guidance considering the end purpose of creating an EPD document. The International EPD System uses GPI 5.0.0 currently as their general instructions. More information is provided on the GPI later in this chapter.

EN 15804:A2 (2021)

Each Programme Operator can have its own PCR for construction products. However, these are based on the European Standard, EN 15804:A2 (2021) with minor adjustments to match their programme. EN 15804:A2 (2021) is part of a series of standards developed by the European Committee for Standardization (CEN) under the Construction Products Regulation (CPR). EN 15804:A2 (2021) is the standard that provides core Product Category Rules (PCRs) for developing EPDs to assess the sustainability of construction products and services. The first version was published in 2012, known as EN 15804:A1 "Sustainability of construction works—Environmental product declarations—Core rules for the product category of construction products". Since 2021, an updated version of the standard has been used within the construction sector, which is called EN 15804:2012 + A2:2019/AC:2021. The current standard includes the rules for calculating the LCI and LCIA for the underlying LCA. In addition, it also includes the guidelines on how to develop an EPD document along with the contents that need to be included in it. The overarching goal of EN 15804:A2 (2021) is to provide transparent and comparable environmental information about construction products [1].

Sub-Product Category Rules

Product Category Rules are a set of specific guidelines that define the procedures for conducting LCA and preparing an EPD for a very specific product within the broader product category; construction products in this case. Not every construction product has a sub-PCR available, and this should be checked within the chosen EPD programme operator before commencing an LCA to ensure compliance.

2.2 The EPD Methodology for Construction Products

The PCR based on EN 15804:A2 (2021) consists of guidelines specific to construction products. By following these guidelines, manufacturers and practitioners of LCA can systematically conduct studies and generate EPDs for construction products that can be communicated accurately and transparently with different stakeholders. Within the context of EN 15804:A2 (2021), the PCR defines the following key methodological choices [4].

Methodological Choices

Methodological choices in an LCA study are critical as they influence the reliability, comprehensiveness, and applicability of the results. Methodological choices encompass various aspects of the LCA process, including goal and scope definition, system boundaries, data collection, impact assessment, interpretation, and reporting. The following section provides a brief description of the different methodological choices with respect to conducting an LCA study for the preparation of an EPD as described in the PCR of EN 15804:A2 (2021).

- System boundary:

The definition of a system boundary is necessary to establish the function of the product system to be considered for the LCA study. The system boundary defines the scope of the study by delineating the processes, activities, inputs, and outputs to be considered throughout the life cycle of the product or system under assessment. According to EN 15804:A2 (2021), system boundaries can be defined through cradle-to-gate, cradle-to-grave, and cradle-to-gate with options, depending on the intended focus of the study.

- Cradle-to-gate: Under this system boundary, only the product phase is included (See Fig. 2, highlighted in orange box).
- Cradle-to-grave: Under this system boundary, the study should encompass all the life phases of the product.
- Cradle-to-gate with options: Under this system boundary, the product phase along with the End-of-Life (EoL) phase of the product is included (See Fig. 2, highlighted in orange and green boxes).

Since the standard is designed from a modular perspective, a product's life cycle can be differentiated into different phases, as shown in Fig. 2. Each phase is further divided into modules, thus enabling the differentiation of the impact originating from different life cycle phases.

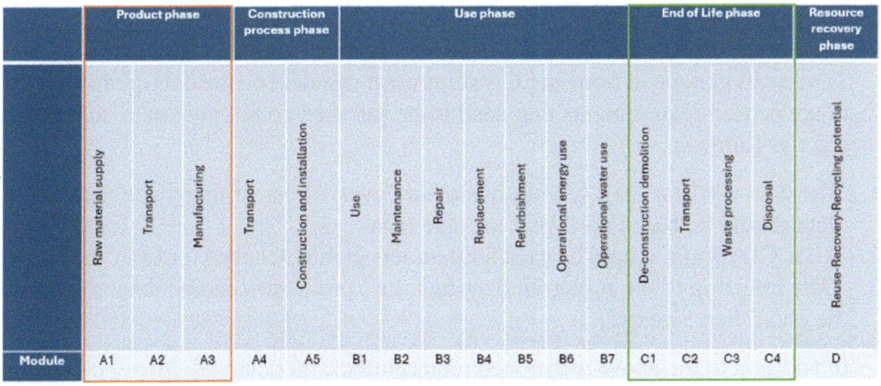

Fig. 2 Life cycle stages of a product according to EN 15804:A2 (2021) [1]

- Module A1—the acquisition of raw materials through extraction and processing, processing of secondary material input (e.g., recycling processes).
- Module A2—the transportation of the raw material to the manufacturer.
- Module A3—the manufacturing process.
- Module A4—the transportation of the product to the construction site.
- Module A5—the installation of the product under study into buildings or roads.
- Module B1-B7—the use stage of a product.
- Module C1-C4—the EoL for products which addresses the activities related to different disposal strategies, recycling, or reuse of the product.
- Module D—benefits or loads beyond the system boundary. Given the EoL scenario used for module C, the product can be assumed to displace the use of a virgin product which serves the same purpose.

- Functional unit and declared unit:

As the construction product can have several applications, a reference unit known as a functional or declared unit needs to be defined. A functional or reference unit provides a reference to which all the material flows, such as inputs and outputs of the manufacturing process, are normalized to express information on a common basis. Depending on the goal and scope of the EPD, a functional or declared unit is used to conduct LCA.

- Functional unit: A functional unit calculates the data for a manufacturing process's inputs and outputs in LCA. Within the framework of EN 15804:A2 (2021), the functional unit shall specify the function, reference service life, and performance of the product under study to conduct an LCA study.
- Declared unit: According to EN 15084:A2 (2021), a declared unit shall be used in the case where a specific function cannot be defined for a product, meaning that the product can be used for several applications in the context of construction works. The declared unit in the EPD shall be one of the following: mass (kg), length (m), area (m2), volume (m3), or an item [1].

- Data quality:

The results generated from an LCA study are dependent on the data quality. Hence, there are certain requirements that need to be satisfied concerning data quality, and they are as follows:

1. The data used for the LCA study should cover 1 year of manufacturing i.e. the data should be based on 1-year average data.
2. An ILCD format should be used for data sets in the LCI used for LCA modelling.
3. Data referring to the foreground system, i.e., producer-specific data, should not be older than 5 years.

In addition to the above mentioned requirements, there are additional mandatory requirements that need to be satisfied, which are described in the standard.

Table 1 Core impact categories that need to be declared according to PCR in EN 15804:A2 (2021)

Impact category	Unit	Method
Global warming potential total, GWP—total	Kg CO_2 eq	IPCC baseline, 100 years, 2013
Global warming potential fossil, GWP—fossil	Kg CO_2 eq	IPCC baseline, 100 years, 2013
Global warming potential biogenic, GWP—biogenic	Kg CO_2 eq	IPCC baseline, 100 years, 2013
Global warming potential land use and land use change, GWP—LULUC	Kg CO_2 eq	IPCC baseline, 100 years, 2013
Depletion potential of the stratospheric ozone layer, ODP	Kg CFC-11 eq	Steady-state ODPs, WMO 2014
Acidification potential, accumulated exceedance, AP	Mol H^+ eq	Accumulated Exceedance, Seppälä et al. 2006, Posch et al., 2008
Eutrophication potential, fraction of nutrients reaching freshwater end compartment, EP-freshwater	Kg P eq	EUTREND model, Struijs et al., 2009b, as implemented in ReCiPe
Eutrophication potential, fraction of nutrients reaching freshwater end compartment, EP-freshwater	Kg $(PO_4)^{3-}$ eq	EUTREND model, Struijs et al., 2009b, as implemented in ReCiPe
Eutrophication potential, accumulated exceedance, EP-terrestrial	Mol N eq	Accumulated Exceedance, Seppälä et al. 2006, Posch et al. 2008
Formation potential of tropospheric ozone, POCP	Kg NMVOC eq	LOTOS-EUROS, Van Zelm et al., 2008, as applied in ReCiPe
Abiotic depletion potential for non-fossil resources, ADP—minerals & metals	Kg Sb eq	CML 2002, Guinée et al., 2002, and van
Abiotic depletion potential for fossil resources, ADP-fossil fuels	MJ, net calorific value	CML 2002, Guinée et al., 2002, and van
Water (user) deprivation potential, deprivation weighted water consumption, WDP	m^3 world eq. deprived	Available WAter REmaining (AWARE), Boulay et al., 2016

- Inventory analysis:

Inventory analysis includes the process of data collection, and guidance provided in EN ISO 14044:2006 should be followed. In the PCR of EN 15804:A2 (2021), the focus is placed on the allocation methods that should be followed (for further information refer to the standard).

- Impact assessment:

In addition, the PCR provides information regarding the different impact categories that need to be quantified in an LCA [1], along with the characterisation

Table 2 Additional impact categories addressing the dust, toxicity and impact on soil that are optional to declare according to according to PCR in EN 15804:A2 [1]

Impact categories	Unit	Method
Potential incidence of disease due to PM emissions, PM	Disease incidence	SETAC-UNEP, Fantke et al. 2016
Potential human exposure efficiency relative to U235, IRP	kBq U235 eq	Human health effect model as developed by Dreicer et al. 1995 updated by Frischknecht et al., 2000
Potential comparative toxic unit for ecosystems, ETP-fw	CTUe	Usetox version 2 until the modified USEtox model is available from EC-JRC
Potential comparative toxic unit for humans (cancer effects), HTP-c	CTUh	Usetox version 2 until the modified USEtox model is available from EC-JRC
Potential comparative toxic unit for humans (non-cancer effects, HTP-nc	CTUh	Usetox version 2 until the modified USEtox model is available from EC-JRC
Potential soil quality index, SQP	na	Soil quality index based on LANCA

factors and methods that need to be used during the LCIA phase to quantify the environmental impacts. Tables 1 and 2, show the core and additional impact categories respectively, that shall be quantified. Tables 3 and 4 show the resource consumption and waste parameters that need to be quantified.

2.3 General Programme Instructions

The General Program Instructions (GPIs) of an EPD program provide guidelines and requirements for the development, verification, and registration of EPDs. These instructions are designed to ensure consistency, transparency, and reliability of the registered EPDs within a specific EPD program. At present there are several EPD programs in operation across Europe, including: EPD international, EPD Norway, IBU (Institut Bauen und Umwelt), INIES, and EPD Belgium among others. It should be noted that the GPIs for each EPD program vary, and different programme operators are more common in certain countries (i.e. EPD Norway is more common in Norway than in France). While specific GPIs may vary between different EPD programs, they typically cover several key aspects:

1. **Compliance with EN 15804:A2 (2021)**: GPIs ensure that EPDs comply with the requirements and guidelines set forth in EN 15804:A2 (2021). This includes adherence to specific methodologies, data quality standards, and reporting formats outlined in the standard.
2. **Scope and Applicability**: GPIs define the scope of the EPD program under EN 15804:A2 (2021), specifying the types of construction products eligible for

Table 3 Resource use parameters addressing the use of primary and secondary resources

Parameter	Unit	Method
Use of renewable primary energy excluding renewable primary energy resources used as raw materials, PERE	MJ, net calorific value	Based on LCI data
Use of renewable primary energy resources used as raw materials, PERM	MJ, net calorific value	Based on LCI data
Total use of renewable primary energy resources, PERT	MJ, net calorific value	Based on LCI data
Use of non-renewable primary energy excluding non-renewable primary energy resources used as raw materials, PENRE	MJ, net calorific value	Based on LCI data
Use of non-renewable primary energy resources used as raw material, PENRM	MJ, net calorific value	Based on LCI data
Total use of non-renewable primary energy resources, PENRT	MJ, net calorific value	Based on LCI data
Use of secondary material, SM	kg	Based on LCI data
Use of renewable secondary fuels, RSF	MJ, net calorific value	Based on LCI data
Use of non-renewable secondary fuels, NRSF	MJ, net calorific value	Based on LCI data
Net use of fresh water, FW	m3	Based on LCI data

Table 4 Categories addressing waste and other external flows

Parameter	Unit	Method
Other environmental information describing waste categories		
Hazardous waste disposed, HWD	kg	Based on LCI data
Non-hazardous waste disposed, NHWD	kg	Based on LCI data
Radioactive waste disposed, RWD	kg	Based on LCI data
Indicators describing external flows		
Components for re-use (CRU)	Kg	Based on LCI data
Materials for recycling (MFR)	Kg	Based on LCI data
Material for energy recovery (MER)	Kg	Based on LCI data
Exported electrical energy (EEE)	MJ	Based on LCI data
Exported thermal energy (EET)	MJ	Based on LCI data

EPD certification. They may outline criteria for determining eligibility based on product category, material composition, or other relevant factors.

3. **Data Requirements and Methodologies**: GPIs provide guidance on the data requirements and methodologies for conducting LCA studies. This includes specifying the types of data to be collected, the functional unit to be used, and the system boundaries to be defined.

4. **Format and Presentation:** GPIs outline how environmental data should be presented, what should be part of the EPD document, and can discuss impact categories that should be included, which is typically those listed in EN 15804:A2 (2021) (See from Tables 1, 2, 3 and 4). That being said, there are programmes like INIES that requires the declaration of other impact indicators in addition to the ones defined in EN 15804:A2 (2021).

5. **Verification and Certification**: GPIs outline procedures for third-party verification, certification, and registration of EPDs on the EPD program. They specify the requirements for independent review, verification, and certification by authorised reviewers to ensure the credibility of EPDs.

2.4 Applications of EPDs

EPDs can be used in Environmental, Social, and Governance (ESG) reporting by providing standardized and transparent information about the environmental performance of products or services. The following provides an overview of how EPDs can be used in ESG reporting:

1. **Environmental Performance Assessment**: ESG reporting frameworks often require companies to disclose their environmental performance. Since EPDs consist of information regarding the environmental impacts of products or services throughout their life cycle, they can be used in ESG reporting, making EPDs a valuable tool for assessing and reporting environmental impacts.

2. **Transparency and Disclosure**: As transparency is a fundamental principle of EPDs that ensures the credibility, reliability, and usability of environmental performance information, companies can use EPDs to demonstrate their commitment to transparency and environmental responsibility by disclosing detailed information about the environmental impacts of their products or services.

3. **Comparative Analysis**: EPDs enable manufacturers to compare the environmental performance of different products or services within the same product category. This comparative analysis allows manufacturers to identify opportunities for improvement and make informed decisions about product design, sourcing, and manufacturing processes. In ESG reporting, manufacturers can use EPDs to demonstrate progress towards sustainability goals and benchmarks.

4. **Stakeholder Engagement**: EPDs provide valuable information for stakeholders, including investors, customers, regulators, and non-governmental organizations (NGOs), who are interested in the environmental performance of products or services. By including EPDs in ESG reporting, companies can enhance

stakeholder engagement and demonstrate their commitment to environmental sustainability.

5. **Application of EPDs in environmental assessment of buildings:** Digital EPDs (as described in ISO standard 22057:2022)) also have an application for assessing the environmental impact of an entire building or construction project with an integration into Building Information Modelling (BIM). The EPD data in digital EPDs is stored in a structured way using an ILCD + EPD data format. This enables incorporating project specific EPDs into third-party BIM software, allowing construction companies to evaluate the selection of different products based on that particular product and supplier specifications from a holistic environmental perspective.

3 EPDs for Construction Products with Nanoparticles

The following section offers guidelines for the development of Environmental Product Declarations (EPDs) specific to construction products that incorporate nanoparticles.

3.1 *Development of an EPD*

Developing an EPD for the construction products with nanoparticles would entail a process similar to that of any construction product such as steel or bricks. The process of developing an EPD is shown in the Fig. 3. The first step to publishing an EPD is to choose an EPD program where the EPD needs to be published. Later steps include reviewing of the PCR and the GPI of the chosen EPD program. This is followed by conducting an LCA study and going through the verification process to verify the results before publishing the EPD.

Fig. 3 The approximated resources needed in the process of publishing an EPD [6]

Which PCR or sub-PCR is applicable depends on the application of the product. Many construction products containing nanoparticles fall under c-PCR-017 Technical-chemical products (for the construction sector). This c-PCR covers product-specific terms for glues, adhesives, screens, plasters, renders, fine smoothing compounds, sealants, primers, mortars, liquid applied membranes and surface treatment products. If not specified, then no sub-PCR needs to be applied and only the PCR for construction products (EN 15804:A2 (2021)) needs to be followed.

Defining a System Boundary

The system boundary of the product life cycle determines the processes to be included or excluded in the LCA. For a construction product with nanoparticles, a generic system boundary can be represented as shown in Fig. 4. As illustrated, the system boundary can be divided into three stages: upstream, foreground, and downstream processes which can be mapped to the modules given in EN 15804:A2. Specification for system boundaries can be specified in c-PCR for specific application. However, these are not defined in any particular PCR or c-PCR for nanoparticles in construction products. A similar approach, as depicted in Fig. 4, is described in the PCR 2023:02 Graphite products. However, this is not a PCR for construction products.

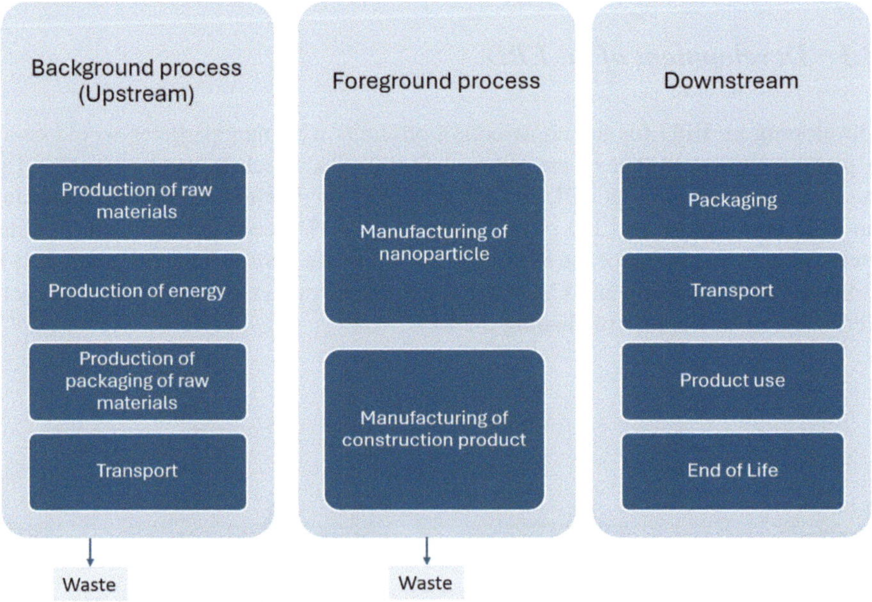

Fig. 4 Illustrating the processes that needs to be included in the product system under assessment, divided into upstream, foreground, and downstream processes. These can then be mapped to one of the 16 sub-modules identified in EN 15804:A2 [1] described in Fig 2.

- **Upstream processes:**

  The upstream process should include the following:

  – Extraction and processing of raw materials required for the core processes,
  – Recycling processes of secondary materials from other product life cycles,
  – Production of input consumables,
  – Treatment of waste generated by the upstream module,
  – Transport of raw materials and components along the upstream supply chain to a distribution point (e.g. a stockroom or warehouse),
  – Production and distribution of packaging,
  – Generation of electricity and production of fuels, steam and other energy carriers used in upstream processes.

- **Foreground processes:**

  The foreground processes, also referred to as 'core processes', include the manufacturing of the nanoparticles and the manufacturing of the studied product, see Fig. 5. In addition to what is shown in Fig. 5, the foreground process also includes:

  – Maintenance of the manufacturing equipment
  – Any emission generated during manufacturing (CO_2, CO, NOx, SOx, heavy metals, PM, etc.). **Note:** Pay attention to double counting.

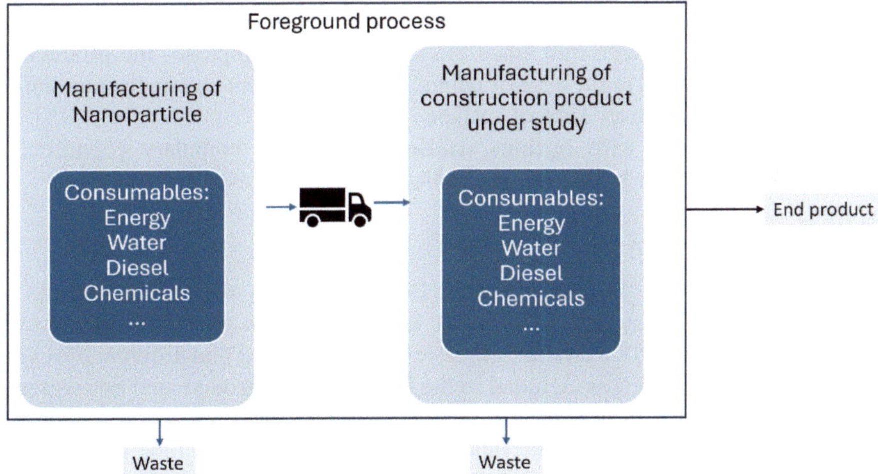

Fig. 5 Foreground process

- EoL treatment of the waste from the manufacturing process, even if it is outsourced to a waste treatment facility outside of the manufacturing facility; in such cases, transportation needs to be included.

- **Downstream processes:**

  The downstream processes of the system boundary include:

  - Transportation to the location where the product is intended to be used (use default scenario when a specific scenario is not defined in a sub-PCR)
  - Product use which refers to the use of any consumables, such as electricity, chemicals, water, etc., needed for the installation, use, and maintenance during the product´s life cycle.
  - EoL refers to the waste treatment process for the used product along with its packaging, including the transportation to the waste treatment facility.

Specifying a Functional/Declared Unit

The definition of the functional or declared unit depends on the system boundary of the assessment. The functional/declared unit for the LCA study can be influenced by the chosen system boundary conditions. Examples of functional units depending on the different system boundaries used can be as follows:

- **In Cradle-to-grave studies**: The functional unit for the LCA study should describe the product's function during its use phase.
- **In Cradle-to-gate studies**: Under this boundary condition, an LCA practitioner can choose the declared unit since the study only encompasses the production phase. For example, in the case of paints, the declared unit can be the amount of product needed to cover 1 m^2 of surface.
- **In Cradle-to-gate with options studies:** Under this boundary condition, a declared unit similar to that of the Cradle to gate can be used.

What to Include in Each Life Cycle Module

As described in section "Defining a System Boundary" the system boundary for LCA can be divided into upstream, foreground, and downstream processes. In contrast, while developing an EPD, a product's life cycle can be divided into different life cycle modules. And the processes included in the upstream, foreground, and downstream processes is included in different life cycle modules according to EN 15804:A2 (2021). The following Fig. 6 and Table 5, describes what to include in each sub-module in the case of construction products with nanoparticles. Data should be collected for one year of production.

Table 5 Description of life cycle modules and the respective data requirements

Life cycle module	What to include	Data requirement
Module A1	**For nanoparticle components of the product:** • Production of raw materials • Production of packaging (if applicable) • Waste generated during the production process • Step by step manufacturing process and the consumables used • Treatment of waste generated during the manufacturing process **For further components of the construction product:** • Production of raw material and packaging • Treatment of waste generated during the production process	• Quantity of consumables such as energy, water, chemicals etc. in production of raw materials (commonly available as data records) • Quantity of waste generated during the process • Quantity of consumables such as energy, water, chemicals etc.in the manufacturing of nanoparticles • Quantity of waste generated during the manufacturing of nanoparticles • Quantity of consumable such as energy, water, chemicals etc • Quantity of waste generated during the process and the respective waste treatment
Module A2	**For Nanoparticle components of the product:** • Transportation distance between the producers of raw materials and manufacturing facility **For further components of the product:** • Transportation distance between the producers of raw materials and manufacturing facility	• Type of transport (eg. Truck, trains and/or ship) • Transportation distance • Quantity of raw material that is transported • Type of transport (eg. Truck, trains and/or ship) • Transportation distance • Quantity of raw material that is transported
Module A3	**For the complete construction product:** • Step by step manufacturing process and the consumables used • Treatment of waste generated during the manufacturing process	• Quantity of different types of consumables used during the manufacturing process • Quantity of waste generated and the respective waste treatment
Module A4	Transportation of final product to the location where it is used	• Type of transport (e.g. Truck, trains and/or ship) • Transportation distance • Quantity of raw material that is transported

(continued)

Table 5 (continued)

Life cycle module	What to include	Data requirement
Module A5	Installation of the product	• Quantity of different types of consumables used during the installation • EoL treatment of packaging waste used for packaging of the final product
Module B1-B7	• B1—Use or application of the installed product • B2—Maintenance of the product • B3—Repair if required during the reference service life (RSL) • B4—Replacement if required during RSL • B5—Refurbishment if required during RSL • B6—operational energy use, if any • B7—operational water use, if any	• Data requirement varies for different products. Hence the data collected here should be specific to product • Data requirement is dependent on which life cycle module form B1 to B7 is included the study
Module C1-C4	• C1—de-construction, demolition • C2—transport to waste processing • C3—waste processing for reuse, recovery and/or recycling • C4—disposal	• Data requirement varies for different products. Hence the data collected here should be specific to product • In case of construction product with nanoparticles, there should specific EoL treatment and hence the data requirement can vary

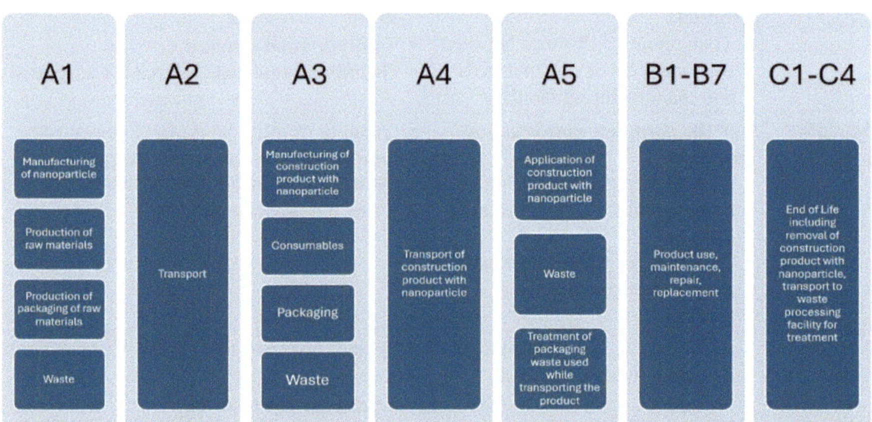

Fig. 6 What to include in each life cycle module. **Note:** Consumables can include electricity, diesel, chemicals, water, and so on. These are the inputs to the manufacturing process that are not part of the final product but are necessary for the production process as auxiliary inputs

3.2 Allocation

Once the data is collected, a complete LCI should be available for one year of production. The data should then be related to the modules and the functional/declared unit (e.g. by using a reference flow). The sum of all the modules and products according to the functional/declared unit should be equal to the sum of the total LCI. Where possible, the division of the modules should reflect real data according to those modules. If this data is missing, the allocation to the separate modules should be described.

If only one product is produced, further allocation is not needed, and the LCI for each module can be directly related to the functional/declared unit. If more than one product is produced, further allocation needs to be considered. EN 15804:A2 (2021) outlines a hierarchy on allocation choices which are described below:

- **Avoid allocation**. Ideally, allocation is avoided in systems producing multiple products by, for example, dividing modules into sub-processes that relate to only one product.
- **Allocation based on physical properties**. If allocation cannot be avoided, then allocation should be conducted based on a relevant physical property of the products, for example, mass.
- **Allocation based on other related descriptions of the product**. If allocation cannot occur based on physical properties, other relationships between the products can be used to allocate the LCI, for example, the economic value of the products [1].

Further guidance is given on allocation in both ISO 14044 and EN 15804:A2 (2021).

3.3 LCIA

After gathering the requisite data for each life cycle module, the subsequent step involves conducting an impact assessment to derive comprehensive LCA results. Following the guidelines outlined in EN 15804:A2 (2021), the impact assessment should adhere to the procedures presented in the standard. Detailed instructions pertaining to this process can be within the EN 15804:A2 (2021) standard documentation.

3.4 Further Steps

Upon completion of the impact assessment, interpretation of the results is needed. To understand the influence of data quality and methodological decisions, a sensitivity analysis should be conducted, and a contribution analysis is also recommended. The results should be used to identify where improvement potentials can exist to reduce the environmental impact of the product. Many of the standards within the EPD framework can provide extra guidance on the interpretation of the results.

After the interpretation of the results and producing a report on the LCA study, the LCA results need to be presented in an EPD. The structure of the EPD may vary depending on the chosen program operator overseeing the EPD's publication. Detailed specifications regarding the EPD structure for a particular program operator can typically be accessed via their respective website.

Subsequently, the generated EPD undergoes a verification process, conducted by an authorized verifier, prior to its publication. This verification process ensures the accuracy, reliability, and adherence to established standards of the EPD. Only upon successful verification does the EPD become available for public dissemination.

References

1. UNE—Asociación Española de Normalización, "UNE-EN 15804:2012+A2:2020/AC:2021. Sustainability of construction works—Environmental product declarations—Core rules for the product category of construction products," 2021. [Online]. Available: https://www.une.org/enc uentra-tu-norma/busca-tu-norma/norma?c=norma-une-en-15804-2012-a2-2020-ac-2021-n00 67399. [Accessed 21 may 2024]
2. International Organization for Standardization (ISO), "ISO 14040:2006. Environmental management — Life cycle assessment — Principles and framework," [Online]. Available: https://www.iso.org/standard/37456.html [Accessed 09 September 2023]
3. International Organization for Standardization (ISO), "ISO 14044:2006. Environmental management — Life cycle assessment — Requirements and guidelines," [Online]. Available: https://www.iso.org/standard/38498.html. [Accessed 09 September 2023]
4. Swedish Environmental Research Institute [IVL], & International EPD System. (2020). PCR 2012:01 Construction products and construction services (EN 15804:A1) (2.33). Available: https://portal.environdec.com/api/api/v1/EPDLibrary/Files/8db1b3cb-e7b9-4716-bd72-c13 9274f5bed/Data [Accessed: 09 September, 2023]
5. M.Z. Hauschild, M.A.J. Huijbregts, in *Introducing Life Cycle Impact Assessment*, eds. by M.Z. Hauschild, M.A.J. Huijbregts. Life Cycle Impact Assessment (Springer Netherlands, Dordrecht, 2015)
6. C. Lee, F. Sánchez, G. Asbjörnsson, M. Evertsson, Application of production simulation combined with site-specific data to quantify the environmental performance of aggregate products at five pilot sites, in *Conference in Minerals Engineering* (2024)

Chapter 9
Interpretation of LCA Results and EPD Comparability

Varun Gowda Palahalli Ramesh and Christina Lee

Abstract The chapter aims to facilitate the interpretation of Life Cycle Assessment (LCA) results and improve the comparability of Environmental Product Declarations (EPDs). This chapter provides guidance on analysing LCA data, grasping fundamental impact categories, and conducting data quality assessments during EPD development. By offering clear methodologies and criteria for assessment, it facilitates the simultaneous comparison of multiple EPDs or individual comparisons, thereby supporting more informed decision-making in the construction industry.

1 Environmental Product Declarations (EPD) Information

The results published in an EPD for any product is a selected sample from a larger and more extensive LCA report usually referred to as the background report which is not publicly available. The purpose of the background report is to convey information associated with conducting the LCA with all assumptions clearly stated, complete LCI results, data records used for modelling, and the allocation of the impact to different products or product groups, among more. Additional analysis is also included in the background report such as contribution and sensitivity analysis. This should enable a certified verifier to verify that the process and results are executed according to the prescribed standards and that the results are plausible.

Present Address:

V. G. P. Ramesh (✉) · C. Lee
Department of Industrial and Material Science, Chalmers University of Technology, Gothenburg, Sweden
e-mail: varung@chalmers.se

C. Lee
e-mail: leec@chalmers.se

© The Author(s) 2025
P. Mercader-Moyano and P. Porras-Pereira (eds.), *Life Cycle Analysis Based on Nanoparticles Applied to the Construction Industry*,
https://doi.org/10.1007/978-3-031-79115-4_9

1.1   EPD Impact Categories

EPD documentation includes essential information that the readers of the EPD requires, for example, information on the definition of system boundaries, which modules are included in the study, definitions of different scenarios, which functional unit is declared, the data quality, and which allocation method is applied. The most valuable information from the EPD is the quantified results of the different impact categories, resource use, waste, and external flows (see Tables 1, 2, 3 and 4 for the included categories and parameters).

Table 1   Core impact categories that need to be declared according to PCR in EN 15804:A2 (2021)

Impact category	Unit
Global warming potential total, GWP—total	kg CO_2 eq
Global warming potential fossil, GWP—fossil	kg CO_2 eq
Global warming potential biogenic, GWP—biogenic	kg CO_2 eq
Global warming potential land use and land use change, GWP—LULUC	kg CO_2 eq
Depletion potential of the stratospheric ozone layer, ODP	kg CFC-11 eq
Acidification potential, Accumulated Exceedance, AP	Mol H^+ eq
Eutrophication potential, fraction of nutrients reaching freshwater end compartment	kg $(PO_4)^{3-}$ eq
Eutrophication potential, fraction of nutrients reaching freshwater end compartment	kg N eq
Eutrophication potential, Accumulated Exceedance, EP-terrestrial	mol N eq
Formation potential of tropospheric ozone, POCP	kg NMVOC eq
Abiotic depletion potential for non-fossil resources, ADP- minerals & metals	kg Sb eq
Abiotic depletion potential for fossil resources, ADP-fossil fuels	MJ, net calorific value
Water (user) deprivation potential, deprivation weighted water consumption, WDP	m^3 world eq. deprived

[*] IPCC—Intergovernmental Panel on Climate Change

Table 2   Additional impact categories addressing the dust, toxicity and impact on soil according to PCR in EN 15804:A2 (2021)

Impact categories	Unit
Potential incidence of disease due to PM emissions, PM	Disease incidence
Potential human exposure efficiency relative to U235, IRP	kBq U235 eq
Potential comparative toxic unit for ecosystems, ETP-fw	CTUe
Potential comparative toxic unit for humans (cancer effects), HTP-c	CTUh
Potential comparative toxic unit for humans (non-cancer effects, HTP-nc	CTUh
Potential soil quality index, SQP	na

Table 3 Resource use parameters addressing the use of primary and secondary resources according to PCR in EN 15804:A2 (2021)

Parameter	Unit
Use of renewable primary energy excluding renewable primary energy resources used as raw materials, PERE	MJ, net calorific value
Use of renewable primary energy resources used as raw materials, PERM	MJ, net calorific value
Total use of renewable primary energy resources, PERT	MJ, net calorific value
Use of non-renewable primary energy excluding non-renewable primary energy resources used as raw materials, PENRE	MJ, net calorific value
Use of non-renewable primary energy resources used as raw material, PENRM	MJ, net calorific value
Total use of non-renewable primary energy resources, PENRT	MJ, net calorific value
Use of secondary material, SM	kg
Use of renewable secondary fuels, RSF	MJ, net calorific value
Use of non-renewable secondary fuels, NRSF	MJ, net calorific value
Net use of fresh water, FW	m3

Table 4 Categories addressing waste and other external flows according to PCR in EN 15804:A2 (2021)

Parameter	Unit
Hazardous waste disposed, HWD	kg
Non-hazardous waste disposed, NHWD	kg
Radioactive waste disposed, RWD	kg
Components for re-use, CRU	Kg
Materials for recycling, MFR	Kg
Material for energy recovery, MER	Kg
Exported electrical energy, EEE	MJ
Exported thermal energy, EET	MJ

Note All EPD results are relative and are only comparable to the same impact category for a similar product applying the same PCR or c-PCR with similar assumptions. The flexibility of the EPD frameworks allows for different assumption based on product/production specific information, different system boundaries, and the use of different databases for the quantification

The following sections emphasize crucial aspects necessary for interpreting EPD information, enabling meaningful comparisons between different EPDs within the same product category.

1.2 Understanding Core Impact Categories

As stated in the PCR of EN 15804:A2 (2021), there are 13 core impact categories that shall be declared in an EPD. It is important to understand what these categories mean and what they measure. Table 5 gives an overview of the core impact categories.

The EPD framework also outlines six additional impact categories which are not discussed in Table 5. This omission is because these categories are optional for disclosure in an EPD as per EN15804:A2 (2021) due to their high uncertainty. Similarly, the categories within resource consumption and waste reported in Tables 3 and 4 respectively, are not elaborated further as they rely directly on the Life Cycle Inventory (LCI) data.

2 Data Quality Assessment

Data quality assessment is an important step in developing the EPDs. This information is also essential for the reader of the EPD to assess the validity of the results. The assessment of data quality is based on the following:

- **Geographical representativeness** : LCA results are sensitive to the geographical representativeness of the data records used while modelling the different processes. For example, consider a hypothetical production process occurring in China that requires data referring to electricity use to calculate the impact. It's necessary that the data record used to model the electricity use accurately reflect the conditions specific to China, for example, the proportions of energy production sources (solar, wind, coal powerplant etc.), otherwise, there is a risk of overestimating or underestimating the environmental impacts.
- **Temporal representativeness:** The data records used for modelling the different processes shall not be older than 5 years in case of processes manufacturing of the product under assessment. In the instances where generic data is used, then the data records shall not be older than 10 years. Refer to Table E.1 in the appendix of EN 15804:A2 (2021) standard for further clarifications.
- **Technological representativeness** : The methods used for modelling the foreground processes shall be representative of the actual manufacturing process for the product and it should represent the state-of-the-art manufacturing process that exists at the manufacturing facility.

Refer to Table E.1 in the appendix of EN 15804:A2 (2021) standard that describes the data quality assessment scheme used to assess generic and specific data records in the LCA study. The assessment scheme is a direct extract from EN15804:A2 (2021) and should be used for EPDs on construction products.

Table 5   Brief overview of core impact categories (the definitions are drawn from sources such as EN 15804:A2 (2021) and EPD-Belgium)

Impact category	Description
Global warming potential total, GWP—total	Global Warming Potential total (GWP-total) which is the sum of GWP-fossil, GWP-biogenic, and GWP-luluc.
Global warming potential fossil, GWP—fossil	The global warming potential related to greenhouse gas (GHG) emissions that originate from the fossil fuels by means of their transformation (e.g. combustion, digestion, etc.).
Global warming potential biogenic, GWP—biogenic	The global warming potential related to carbon emissions to air (CO_2, CO and CH_4) originating from the oxidation and/or reduction of aboveground biomass through transformation or degradation (e.g., combustion, digestion, composting, landfilling) and CO_2 uptake from the atmosphere through photosynthesis during biomass growth—i.e. corresponding to the carbon content of products, biofuels, or above ground plant residues such as litter and dead wood.
Global warming potential land use and land use change, GWP—LULUC	The global warming potential related to carbon uptakes and emissions (CO_2, CO and CH_4) originating from changes in carbon stock caused by land use and land use change. This sub-category includes biogenic carbon exchanges from deforestation, road construction or other soil activities (including soil carbon emissions).
Depletion potential of the stratospheric ozone layer, ODP	Measures the impact on stratospheric ozone layer caused by the breakdown of certain chlorine and/or bromine-containing compounds (chlorofluorocarbons or halons). These compounds break down and catalytically destroy ozone molecules.
Acidification potential, accumulated exceedance, AP	Measure the impact of acid depositions on the environment. The main sources for emissions such as SO_2, NOx, and NH_3 are agriculture and fossil fuel combustion.
Eutrophication potential, fraction of nutrients reaching freshwater end compartment, EP-freshwater	The potential to cause over-fertilization of freshwater as a result of increased growth of algae in fresh water and the following impacts.
Eutrophication potential, fraction of nutrients reaching freshwater end compartment	The potential to cause over-fertilization of marine water, which can result in increased growth of biomass.

(continued)

Table 5 (continued)

Impact category	Description
	This is focused on waterborne and airborne nitrogen emissions.
Eutrophication potential, accumulated Exceedance, EP-terrestrial	The potential to cause over-fertilization of soil, which can result in increased growth of biomass and following impacts.
Formation potential of tropospheric ozone, POCP	The potential to create ground-level ozone which is harmful to organisms. This is caused by chemical reactions brought about by the light energy of the sun creating photochemical smog.
Abiotic depletion potential for non-fossil resources, ADP- minerals & metals*	Consumption of non-renewable resources, their availability for future generations.
Abiotic depletion potential for fossil resources, ADP-fossil fuels*	Measure for the depletion of fossil fuels such as oil, natural gas, and coal. The stock of the fossil fuels is formed by the total amount of fossil fuels.
Water (user) deprivation potential, deprivation weighted water consumption, WDP*	Accounts for water use related to the local scarcity of water as freshwater is a scarce resource in some regions, while in others it is not.

*The results of this environmental impact indicator shall be used with care as the uncertainties on these results are high or as there is limited experience with the indicator

Besides considering the geographical, technical, and temporal representation, it is important to systematically address aspects of precision, completeness, representativeness, consistency, and reproducibility to ensure the validity of EPD results. Precision demands the data reflects the environmental impacts with minimal variability, whereas completeness involves covering all environmental flows from resource extraction to End-of-Life (EoL) (Note: This can vary depending on the system boundary). Representativeness ensures the data aligns closely with the actual conditions of the product's lifecycle, including geographical location, technology used, and relevant time frames. Consistency across the study guarantees that methodologies are applied uniformly, facilitating reliable comparisons between different products. Reproducibility emphasizes the importance of documenting the assessment process comprehensively, so independent practitioners can replicate the results. Moreover, specifying all sources of data is crucial, whether originating from datasets, models, or assumptions and expressing the uncertainty of information clearly, acknowledging any potential variability or assumptions made during the assessment. These elements collectively define the data quality.

3   EPD Programme Operators

The initial step in creating an EPD is to select an appropriate EPD program operator which are often regional. Several EPD program operators exist today whose purpose is to oversee the verification and publication of EPDs. At present, there are at least 18 EPD program operators that are in operation [7]. Table 7 provides an overview of a few common EPD program operators in Europe.

Rules and guidelines from different EPD programme operators are not completely harmonized. Occasionally, EPD program operators such as IBU supplement the core PCRs defined in EN 15804:A2 (2021) with c-PCRs, also called Part B PCRs. A Part B PCR provides additional rules and guidelines for a specific group of products

Table 7 List of EPD program operators

EPD Operator	Description
EPD International [2]	EPD International oversees the global EPD system with the headquarters located in Sweden, providing a framework for the development and verification of EPDs across various industries and regions worldwide. Their standards ensure consistency and comparability of EPDs on a global scale. The program operates in accordance with ISO 14025, ISO 14040, ISO 14044, and EN 15804:A2 (2021). The International EPD System has a global service network with exclusive representations in countries such as Argentina, Australia, Bangladesh, Brazil, Chile, Egypt, India, Mexico, New Zealand, Russia, Southeast Asia (Indonesia, Malaysia, the Philippines, Singapore and Vietnam) and Turkey.
IBU (Institut bauen und umwelt e.V.) [3]	IBU is a German EPD program operator primarily focused on construction and building materials. They provide EPD services for various construction products in compliance with European standards such as EN 15084:A2. (2021).
EPD-Norway [1]	EPD-Norge is the Norwegian EPD program operator responsible for managing EPDs in Norway. The program allows the publication of EPDs which accordance with ISO 14025, ISO 21930, and EN 15804:A2(2021) on their platform.
EPD Denmark [4]	EPD Danmark serves as the operator of the Danish EPD program, tasked with managing Environmental Product Declarations (EPDs) in Denmark. Their responsibilities include overseeing the verification and publication of EPDs for construction products in compliance with ISO 14025 and EN 15804:A2 (2021).
EPD Ireland [5]	EPD Ireland is an EPD program operator focusing on Ireland. They manage the EPD process for various products and sectors within the Irish market, ensuring compliance with relevant standards and guidelines.
EPD Belgium [6]	EPD Belgium is an EPD program operator focusing on the Belgian market. The main aim of the Belgian EPD program is to support sustainable construction and procurement practices in the Belgian market. B-EPD issues EPDs for construction products that comply with ISO 14025 and EN 15804:A2 (2021).

like specific construction materials. The program operators can issue these Part B PCRs independently which has led to inconsistencies while developing EPDs for the same product [7, 8]. In addition, there are some differences due to different stakeholders being involved in setting these rules and guidelines on a product level. These can be methodological differences such as cut-off rules, modelling approaches, or allocation rules. Although EN 15804:A2 (2021) has been successful to a certain extent in harmonizing the EPDs of the same product category, a study by Gelowitz and McArthur [8] finds that 3–12% of the EPDs are incomparable even while following the same PCR, and 73–87% of the EPDs for the same product category following different PCRs are incomparable.

However, there are some mutual recognition agreements in place. The EPD operators in Sweden, Denmark, and Norway have an agreement on mutual recognition of EPDs, including the PCR and c-PCR they operate under. This implies that owners of EPDs, namely manufacturers, will gain significantly different visibility and access to a broader market. Concurrently, it will simplify the process for consultants, contractors, and others involved in calculating the climate impact of buildings to obtain data on the building products employed in their construction projects.

4 Comparability of EPDs

When comparing EPDs there is an option to compare multiple EPDs at the same time or comparing one-on-one. In most cases, one-on-one comparison occurs. Comparing multiple EPDs at once is labour intensive as compiling information from multiple EPDs for statistical analysis is still a very manual process. This will hopefully improve as digital EPDs become more readily available. It can, however, be highly beneficial for the creator of an EPD to get information on how they stack up to the rest of the industry. Comparing an individual EPD to another is more common for customers receiving the EPDs in business-to-business communication and deciding on a better option based on multiple criteria.

Statistical Analysis

Benchmarking[1] product impacts from EPDs is a challenging task for multiple reasons, as discussed in the previous section. There are several factors influencing the comparability of different EPDs. These are, for example, based on the selection of the Programme Operator, the PCR or c-PCR, data used for the quantification, involved databases, and how the data collection has been performed, to name a few. On top of sources of uncertainty from the EPD process, there is also an uncertainty in the benchmarking as well: how well do the available EPDs represent the overall market?

---

[1] Benchmarking is a systematic approach for comparing business process and performance metrics to the industry performance distribution.

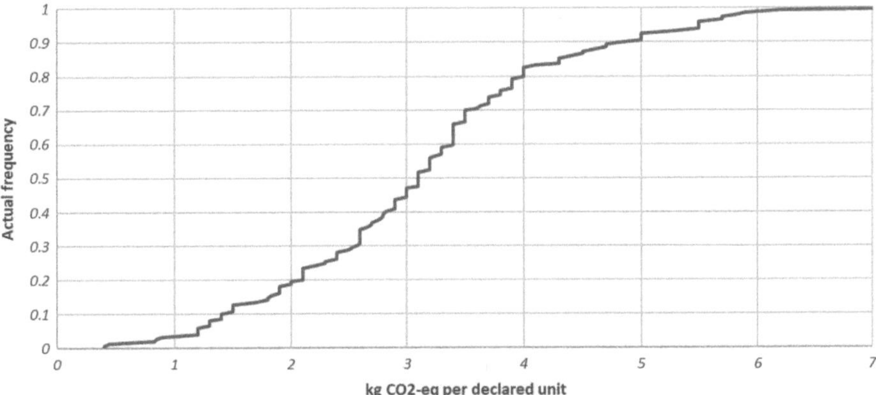

Fig. 1 Cumulative distribution for GWP contribution for aggregates

The number of EPDs is continuously increasing and for some products, there are enough for statistical analysis. For example, for aggregates in Sweden, there are EPDs for 148 product groups from 28 quarries. This enables us to get a representative distribution curve of the impacts. Figure 1 illustrates the cumulative distribution of GWP of aggregates in Sweden. The number of similar construction products containing nanoparticles are still too low to perform similar benchmarking to the industry distribution, as can be done for aggregates. With the introduction of digital EPDs though, the compilation of information could be automated and kept up to date for even smaller product sample sizes.

The shape of the distribution curve in Fig. 1 is close to a normal distribution curve, and other products follow a similar pattern with a narrow and steep inclination centrally. In these cases, the mean and median values are of a similar value. However, the distribution curve is not always normally distributed. If there is a significant difference in the mean and median, the distribution can be skewed or even bimodular, especially if you are evaluating multiple impact categories. The mean and median of the distribution of GWP for aggregates in Sweden is 3.13 kg CO_2 eq. per tonne and 3.1 kg CO_2 eq. per tonne respectively. This indicates a close to normally distributed sample. With the collected product EPD results from multiple EPDS at hand, producers could compare their performance with the overall industry performance. If aggregate producers have a product that has GWP impact of 2 kg CO_2 eq. per tonne then they would be able to assess that they were outperforming 80% of the industry's product groups. The same approach could be done on all impact categories to evaluate the overall impact of the product. This can help individual manufacturers determine if major improvements could be achieved in their production process from the perspective of its environmental performance, and encourage competition for more environmentally friendly manufacturing processes. For an additional example on the impact distributions, see the Welling and Ryding study on insulation material [9].

Comparative Analysis

While comparing the results for different impact categories in EPDs for the same product from different manufacturers, it is essential to understand the differences in methodological choices, databases used, cut-off criteria, allocation methods and the assumptions in the LCA framework. This is because these aspects will influence the overall results of an LCA study. The following section provides a brief description of how the differences in different methodological choices and databases used for the LCA study will influence the results.

- Differences in Functional or Declared Unit

An EPD for a construction product can be published by using either a functional (FU) or declared unit (DU) (Part 4, Chap. 8, Sect. 2.2 for further information). Although the LCA standards, ISO14040 and ISO14044, specify that a FU should be used while conducting an LCA, the core PCR in EN15804:A2 (2021) allows for the use of a DU in the place of a FU. For example, a DU can be 1 kg of paint without reference to the function it performs. A DU is recommended to be used when a product has multiple functions, or the function is unknown. Using a DU usually results in the exclusion of the use phase (module B1-B7) from the study. Choosing DU or FU while conducting an LCA study influences the system boundary definition, along with the input process that is included in the study. Therefore, a comprehensive understanding of both the DU and FU is imperative while comparing the EPDs of a construction product under the same product category.

- Differences in System Boundary

As described in earlier chapters, defining a system boundary is dependent on the goal of an LCA study. In the context of an EPD, the inclusion or exclusion of a life cycle module from the assessment depends on the scope, the choice between FU or DU, and the availability of data. The core PCR allows for publishing EPDs with different system boundaries (Refer to the system boundary section in Part 4, Chap. 8, Sect. 2.2). Consequently, EPDs for a construction product under the same product category manufactured by different manufacturers can only be compared across the same system boundary. The reason is that the system boundary defines the unit processes to be included in the assessment. This influences the LCA results reported in the EPD. For example, consider the EPDs for a steel structure manufactured by two different manufacturers located in the same area. The EPD from the first manufacturer covers the product phase (A1-A3) and the EoL phase (C1-C4), and the EPD of the second manufacturer covers all life cycle phases (A1-A3, A4, A5, B1-B7, C1-C4). In this case, these EPDs cannot be directly compared due to the differences in system boundary.

- Differences in Databases

EPDs for construction products can be created using data from different databases. Using manufacturer-specific data is important while developing an EPD though. To develop an LCA, manufacturer-specific data (i.e., Foreground data) is collected, which can then be modelled by using data records from different databases such as Ecoinvent, GaBi, Ökobaudat, and others to provide information on the environmental impact. Taking electricity as an example, the data record used for the modelling of the electricity should be representative of the national electricity production where the product is manufactured. In the context of EPDs, this can vary depending on the EPD program operators. For instance, in order to model electricity consumption during the manufacturing of a certain construction product in LCA, the GPI of different program operators, for example IBU and international EPD system, allow for the use of both Ecoinvent and GaBi, however, in the case of EPD Norway, Ecoinvent should be used. Using different databases to assess the same product can lead to differences in the results. The difference in results can depend on several factors such as the version of the database and quality of the data record.

In addition to differences in FU or DU units, system boundaries, and databases used in an LCA study to develop an EPD, there are other factors influencing the comparability of the product. These include allocation methods, assumptions regarding transportation, and more. One more aspect that could specifically be influenced by the integration of nanoparticles into the product is the Reference Service Life (RSL) since in most application the purpose of integrating nanoparticles is to improve the product longevity. However, this should not be included in the EPD according to c-PCR-017 *Technical-chemical products (for construction sector)*. Therefore, when interpreting and comparing EPDs for the same product from different manufacturers, it's crucial to consider all these aforementioned factors and be aware of the GPI, PCR and c-PCR involved. This comprehensive approach ensures a more accurate and meaningful comparison of environmental impacts between products, facilitating informed decision-making.

5 Sensitivity Analysis in LCA

In LCA studies, sensitivity analysis explores possible future scenarios based on manufacturer-specific assumptions. A sensitivity analysis can be used to assess the sensitivity of the LCA results. The scenarios described in this section provide insight into multiple ways of conducting a sensitivity analysis in an LCA study. The scenario is often chosen to provide insights to the manufacturer about the possible future. Hence the sensitivity analysis described under this section is directed towards the manufacturer of the product.

Sensitivity analysis can be performed in multiple ways, for example:

1. Sensitivity analysis based on the quantity of the input raw material and quality of the data record used for modelling.
2. Sensitivity analysis based on the transportation to the installation site.
3. Sensitivity analysis based on EoL treatment.
4. Sensitivity analysis based on the allocation method.
5. Sensitivity analysis based on changes in process configuration.

Table 8 Comparison of different scenarios covering A1-A3

Core impact categories	Baseline (total for A1-A3 per DU for a product)	Scenario—increase in significant input by X %	Scenario—using lower quality data
GWP-total	–	–	–
GWP-fossil	–	–	–
GWP-biogenic	–	–	–
GWP-LULUC	–	–	–
GWP-GHG	–	–	–
ODP	–	–	–
AP	–	–	–
EP-freshwater	–	–	–

The following section provides a brief description of the first three examples. Steel as a product will be used as an example to illustrate the different possibilities for sensitivity analysis.

Note: Within the EPD framework the scenarios are usually described in detail in the LCA background report. Only a brief description and the results of these scenarios is later reported in the EPD, if at all, depending on the scope of the EPD.

Sensitivity Analysis Based on Raw Material Quantity

Let's consider a manufacturing process to produce structural steel components. For such a process, raw material such as iron ore, ferrosilicon, ferromanganese, ferrochromium, and ferrovanadium are required in different quantities to manufacture a certain quality of steel. In a baseline[2] assessment, the LCA model is developed from real data from the manufacturer to assess the environmental impact of the steel structure. To explore a possible scenario where the quantity of one of the raw materials needs to be increased by "X %" to manufacture a different quality of steel, a sensitivity analysis can be conducted. The influence of such change can be modelled in LCA to assess its influence on the environmental impact of the product by increasing the material flow of the raw material for "X %" in a hypothetical model. This scenario can be expanded to include the influence of data quality on the LCA results. For example, a lower-quality data record can be used to calculate LCA results and compare them with the results from the baseline assessment.

Table 8 is an example of how such results from a sensitivity analysis covering modules A1—A3 could be presented.

Sensitivity Analysis Based on the Transportation to Manufacturing Site

In industries like steel manufacturing, transport distance plays a significant role in the overall assessment. Take, for instance, the shift towards circularity in steel production, where 'green iron' is now being utilized. This transition is motivated by

---

[2] Baseline refer to the LCA model for the actual manufacturing process NOT the scenario.

Table 9 Comparison of different scenarios covering transportation

Core impact categories	Scenario 1—baseline results per DU for steel using conventional iron with a default distance of 100km	Scenario 2—baseline results per DU for steel using green iron with a default distance of 100 km	Scenario 3—baseline results per DU for steel using green iron with a distance of 2000 km
GWP-total	–	–	–
GWP-fossil	–	–	–
GWP-biogenic	–	–	–
GWP-LULUC	–	–	–
GWP-GHG	–	–	–
ODP	–	–	–
AP	–	–	–
EP-freshwater	–	–	–

Note An EoL route need not be one or the other. Depending on the type and complexity of a product, an EoL treatment route can be a combination of reuse and recycling. In the case of steel, a hybrid EoL route can be that a portion of steel scrap is subjected to reuse and the remainder is recycled

its lower environmental impact compared with steel manufactured using traditional processes, as it removes fossil fuels from the manufacturing process, replacing it with green hydrogen [10].

Here is where the transport distance and transport mode play a pivotal role. Hence conducting a sensitivity analysis is necessary to effectively compare the two manufacturing routes—one utilizing conventional iron and the other relying on green iron. In the case of green iron, the distance between the steel manufacturer and the green iron producer could be much larger and if the transport distance is significant, it could potentially offset the environmental benefits gained from avoiding conventional iron. Hence conducting a sensitivity analysis could be beneficial to understand the trade-offs.

Table 9 shows an example of how scenarios encompassing the influence of changes in transport distance and change in the mode of transportation[3] could be presented. This scenario can be expanded to include the influence of changes in transportation mode on the LCA results.

Sensitivity Analysis Based on EoL Treatment

By examining various scenarios, such as different disposal methods or recycling options, one can better understand the environmental impacts of products throughout their entire life cycle.

Considering a steel product as an example within the cradle-to-grave system boundary of LCA, it becomes essential to include the EoL treatment stage when the steel product reaches its End of Waste (EoW) state. For instance, the steel product

---

[3] Mode of transportation refers to the different transportation modes such as truck, train, ship etc.

Table 10 Example for presenting the sensitivity analysis for EoL treatment

Core impact categories	Results for C1-C4 per DU for a product			
	Scenario—reuse	Scenario—recycle	Scenario—combined reuse and recycling	Scenario—landfilling
GWP-total	–	–	–	–
GWP-fossil	–	–	–	–
GWP-biogenic	–	–	–	–
GWP-LULUC	–	–	–	–
GWP-GHG	–	–	–	–
ODP	–	–	–	–
AP	–	–	–	–
EP-freshwater	–	–	–	–

at its EoL can either be reused without additional treatment if it meets quality standards, or it can undergo treatment as scrap steel and enter a recycling process. In such cases conducting thorough sensitivity analyses can be beneficial. By exploring various EoL treatment routes, such as reuse, recycling, or disposal, a manufacturer can gain insights into how each option affects the steel product's environmental profile throughout its life cycle.

In conclusion, a sensitivity analysis can encompass the whole life cycle of a product or just a portion of it. It mainly depends on the goal and scope of the LCA study and also the availability of data to conduct a robust scenario analysis. For producers of construction material that could potentially incorporate nanoparticles, they could evaluate the performance per DU of including nanoparticles in their products, evaluate different quantities, look at different suppliers or set up different scenarios based on EoL associated with different strategies.

References

1. Epd-norway, [Online]. Available: https://www.epd-norge.no/. (2022). [Accessed 14 May 2024]
2. The International EPD System, Global EPD programme for publication of ISO 14025 and EN 15804 compliant EPDs, [Online]. Available: https://www.environdec.com/about-us/the-international-epd-system-about-the-system. [Accessed 14 May 2024]
3. IBU, About EPD Denmark. Retrieved 09/12/2024 from https://ibu-epd.com/en/ (2024)
4. EPD Denmark, About EPD Denmark, [Online]. Available: https://www.epddanmark.dk/om-epd-danmark/. (2022). [Accessed 14 May 2024]
5. EPD Ireland, Develop an EPD, [Online]. Available: https://www.igbc.ie/develop-an-epd/. (2023). [Accessed 14 May 2024]
6. B-EPD program operator, The Belgian EPD programme B-EPD, [Online]. Available: https://www.health.belgium.be/en/belgian-epd-programme-b-epd. (2022). [Accessed 14 May 2024]
7. F. Konradsen, K. Hansen, A. Ghose, M. Pizzol, Same product, different score: how methodological differences affect EPD results. Int. J. Life Cycle Assess. **29**, 291–307 (2024)

8. M. Gelowitz, J. McArthur, Comparison of type III environmental product declarations for construction products: material sourcing and harmonization evaluation. J. Clean. Prod. **157**, 125–133 (2017)
9. S. Welling, S. Ryding, Distribution of environmental performance in life cycle assessments—implications for environmental benchmarking. Int. J. Life Cycle Assess. **26**, 275–289 (2021)
10. Stegra, Green iron. https://stegra.com/green-iron. [Accessed 2024]

Chapter 10
Results and Discussion

Paula Porras-Pereira and Pilar Mercader-Moyano

Abstract This chapter presents a practical application of LCA within the context of nanotechnology in construction. Specifically, it focuses on an environmental evaluation aimed at determining the more sustainable alternative between a nano-silica-modified asphalt mixture and an unmodified asphalt mixture used in road paving. The practical exercise, titled Comparative Life Cycle Assessment (LCA) of Unmodified and Nano-Silica-Modified Asphalt Mixtures for Road Paving: A Practical Exercise, provides a detailed comparison of the two materials, emphasizing their environmental implications across key life cycle stages. By integrating nanotechnology into conventional construction materials, this study contributes to the understanding of its potential environmental benefits and challenges, offering valuable insights for sustainable infrastructure development.

Keywords Life Cycle Assessment · Nanomaterials · Construction · Sustainability · Nanosilica · Asphalt

After the investigation, it is concluded that one of the fundamental methodologies for assessing the environmental ramifications of a product or service is through the application of Life-Cycle Assessment (LCA). In this section, we delve into a practical demonstration of conducting a Life Cycle Assessment within a nanotechnology construction context, which main objective is to conduct an environmental evaluation to discern the most ecologically sound option between the utilization of a nano-silica-modified asphalt mixture and an unmodified asphalt mixture on the process of paving a road. The practical exercise is titled:

Present Address:
P. Porras-Pereira (✉) · P. Mercader-Moyano
Department of Building Construction I, Higher Technical School of Architecture, University of Seville, Seville, Spain
e-mail: pporras@us.es

P. Mercader-Moyano
e-mail: pmm@us.es

© The Author(s) 2025
P. Mercader-Moyano and P. Porras-Pereira (eds.), *Life Cycle Analysis Based on Nanoparticles Applied to the Construction Industry*,
https://doi.org/10.1007/978-3-031-79115-4_10

Comparative Life Cycle Assessment (LCA) of Unmodified and Nano-Silica-Modified Asphalt Mixtures for Road Paving: A Practical Exercise

Asphalt stands as the predominant choice for pavement across the globe, boasting unparalleled popularity. It is comprised of bitumen, serving as the binding agent, intertwined with either crushed or natural aggregates. This amalgamation of components culminates in the creation of asphalt mixtures, the cornerstone of modern road construction. Consequently, asphalt mixtures are endowed with the capacity to offer unparalleled driving comfort while also facilitating adaptable maintenance procedures. The design of asphalt pavements is meticulously crafted to ensure peak performance over their intended lifespan. Nevertheless, these pavements undergo deformations within relatively short timeframes. This, exacerbated by escalating traffic volumes and harsh weather phenomena, has prompted authorities overseeing asphalt pavements to explore alternative remedies aimed at bolstering their resilience against mechanical wear and environmental stressors [1].

In this context, although few additives exist, nanotechnology and nanomaterials have recently garnered considerable attention from the pavement industry. Particularly, the utilization of nanomaterials as modifiers for asphalt is experiencing a surge in popularity, owing to their distinctive properties that significantly enhance asphalt binder performance [1].

Incorporating nano-silica into asphalt mixtures has been demonstrated to yield numerous advantages. It enhances resistance to oxidative aging, mitigates rutting effects, improves rheological properties, reduces asphalt molecule interaction, thereby lowering mixing and compaction temperatures. Furthermore, it bolsters adhesion, resilience against cracking and rutting, albeit at the expense of decreased ductility and temperature sensitivity [1].

Despite the manifold benefits of integrating nanomaterials into asphalt, uncertainties loom over their environmental repercussions. Hence, as aforementioned, this study endeavours to evaluate nano-silica-modified and unmodified asphalt mixtures using the LCA methodology.

To accomplish the main objective of the practical demonstration of the investigation, initially, an LCA for an unmodified asphalt mixture will be conducted. Subsequently, the results will be thoroughly analysed and contrasted, based on their environmental impact categories, with the provided LCA results for a nano-silica-modified asphalt mixture. Ultimately, the most environmentally preferable asphalt variant will be identified. The application of LCA to NMAM offers a valuable tool for informing stakeholders and decision-makers in the judicious selection of pavement modification additives, empowering them to harness the complete advantages of integrating nanomaterials into pavements, while concurrently mitigating any possible adverse environmental impacts.

1   Life Cycle Assessment of an Unmodified Asphalt Mixture

In order to carry the LCA for an unmodified asphalt mixture, it is essential to define the construction process of paving a street, which encompasses a series of intricate procedures aimed at guaranteeing the longevity, safety, and visual appeal of the roadway. To execute the paving of a new road correctly, multiple layers must be laid, each serving.

In this instance, the following layers will be implemented on the existing compacted natural ground: a sub-base layer, a base layer, an intermediate layer, and a surfacing layer. Below, the various components comprising the construction detail are elaborated upon (Fig. 1).

Considering the preceding statements, an analysis is conducted to determine the LCA of an unmodified asphalt mixture with the aim of identifying and quantifying the materials utilized. As aforementioned, the Functional Unit (FU) serves as the foundation of all LCA investigations. It represents a quantified measure of a product system's performance, utilized as a standard reference unit in LCA analyses. Establishing a fixed value for the FU is essential, as the resulting environmental impact outcomes across various impact categories are contingent upon this chosen unit. For this study, a Functional Unit of 1000 kg (1t) of asphalt mixtures production was adopted. This assessment focuses initially on segmenting the mixture into its constituent layers, as depicted in Fig. 1. In the initial phase, the construction unit of the layer is determined, delineated by the material constituting the layer itself, which, in this instance, pertains to granitic artificial aggregate.

The Life Cycle Inventory (LCI) phase entails the collection of actual data and the modelling of the product system. In order to gather data on material extraction, processing and production, the evaluation version (2024.f) of the 'Generador de precios de la construcción', a downloadable software tool developed by the Spanish company CYPE Ingenieros, S.A., has been utilized. This software is available for download at https://generadorprecios.cype.es/descarga_generador.htm?0.

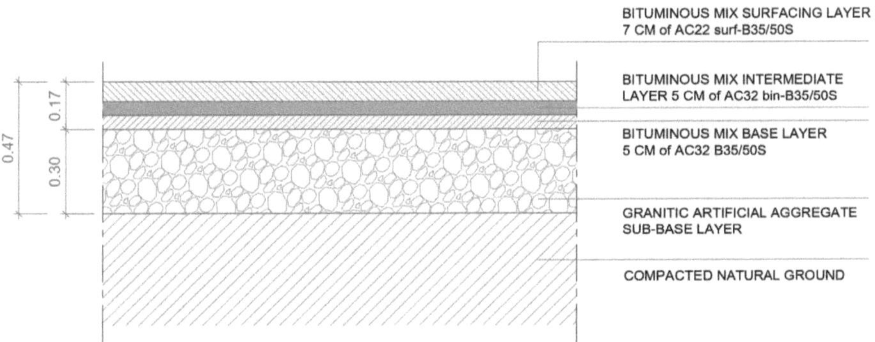

Fig. 1 Road pavement construction section of an unmodified asphalt mixture

Following this, the Life Cycle Impact Assessment (LCIA) phase ensues. During this stage, the collected data undergoes analysis to assess its contribution to each impact category. LCIA encompasses characterization, normalization, evaluation, and weighting, depending on the specific LCIA methodology utilized.

The initiation of the LCA process of an unmodified asphalt mixture, the 'Construction Price Generator' software will be accessed, which should have been previously downloaded. Upon program initiation, the tab for project type and geographical location selection will be opened. For this street paving exercise, 'urban spaces' will be chosen as the project type, with Spain designated as the location. Subsequently, within the left-hand side brown menu of the interface, the precise work location will be defined by selecting the province followed by the municipal area.

- Sub-base layer

To initiate the LCA process, the first step involves evaluating the impact of the sub-base layer. Subsequently, in the broken-down prices tab of the menu, the material constituting the sub-base layer to be analysed, artificial granitic gravel, will be searched for. In this case, it can be found under the categories of urban pavements and pavements—bases and subbases—granular—granular subbase. Once the material is located, its properties will be configured, including the filling material (artificial granite gravel) and the compaction level relative to the maximum dry density achieved in the Modified Proctor test, set at 95%. With the material parameters defined, its information, including breakdown price, specifications, generated waste, environmental impact indicators, and health and safety data, can be extracted. For this instance, data concerning the breakdown price, waste generated, and environmental impact indicators will solely be retrieved using the export option in PDF format situated on the right-hand side of the interface. Following that, the downloaded information will be analysed.

In the following table (Table 1), which identifies and quantifies the materials utilized for the layer construction, a detailed description of the material of the respective layer is provided first, along with its identification code and unit of measurement. Additionally, in the second part of the table, divided into six columns, the different materials comprising the layer are categorized, along with the equipment, machinery, and labour required for its execution, as well as the complementary direct costs stemming from it. The "code" column represents the identification code of the corresponding material used by the software tool employed; the "unit" column denotes the unit of measurement for the material; the "description" column offers a brief description of the material; the "performance" column indicates the execution time in hours, based on the equipment, machinery, and labour utilized; the "unit price" column defines the hourly rate of execution; and finally, the "cost" column represents the actual cost of execution, taking into account the performance and unit price.

Table 1 Unit cost of the sub-base layer

MBG010	m^3	Granular base			

Granular base with artificial granite gravel, and compaction at 95% of the Modified Proctor with mechanical means, in 30 cm thick layers, until reaching a dry density of no less than 95% of the Modified Proctor of the maximum obtained in the Modified Proctor test, carried out according to UNE 103.501, to improve the resistant properties of the soil. The price does not include the performance of the Modified Proctor test

Code	Unit	Description	Performance	Unit price	Cost
1	*Materials*				
mt01zah010d	t	Artificial granite gravel	2.200	10.93	24.05
				Materials subtotal	24.05
2	*Equipment and machinery*				
mq02rot030b	h	Self-propelled tandem compactor, 63 kW, 9.65, working width 168 cm	0.110	46.83	5.15
mq04dua020b	h	Front dump dumper with 2 t payload	0.110	10.58	1.16
mq02cia020j	h	Tanker truck, 8 m^3 capacity	0.011	121.25	1.33
			Equipment and machinery subtotal		7.64
3	*Labour*				
mo113	h	Ordinary construction labourer	0.199	21.19	4.22
			Labour subtotal		4.22
4		Additional direct costs			
	%	Additional direct costs	2.000	35.91	0.72
			Direct costs (1 + 2 + 3 + 4)		36.63

Table 2 Waste generated by the sub-base layer

LER code	Type	Weight (kg)	Volume (l)
01 04 08	Gravel and crushed rock waste other than those mentioned in code 01 04 07	15.603	10.402
	Waste generated	15.603	10.402

To assess waste generation, Table 2 provides detailed information on the waste generated from the execution of the layer, quantified in both mass (kg) and volume (l) units. As expected, since the sub-base layer consists solely of artificial granite aggregate, it constitutes the sole type of waste.

Table 3 Environmental impact and resource use indicators of the sub-base layer

Life cycle stage	Environmental impact indicators							Use of resources		
	GWP	ODP	AP	EP	POCP	ADPE	ADFP	PERT	PERNRT	FW
	CO_2 eq. (kg)	CFC 11 eq. (kg)	SO_2 eq. (kg)	$(PO_4)^{3-}$ eq. (kg)	etileno eq. (kg)	Sb eq. (kg)	(MJ)	(MJ)	(MJ)	(m^3)
Total A1-A2-A3	15.400	2.20e−06	0.220	0.044	0.220	1.10e−05	264.000	15.400	264.000	0.660
A4	2.891	0.004	0.202	0.040	0.011	0.002	318.035		39.071	0.607
A5	2.664	1.76e−07	0.012	0.048	0.002	9.86e−05	37.299		35.878	5.328
Total A4-A5	5.555	0.004	0.214	0.088	0.013	0.002	355.334		74.949	5.936
Total energy and emissions	20,955	0.004	0.434	0.132	0.233	0.002	619.334	15.400	338.949	6.596

A1: Raw materials supply
A2: Transportation of raw materials
A3: Product manufacturing
A4: Product transportation
A5: Construction and installation process
GWP: Global warming potential
ODP: Stratospheric ozone depletion potential
AP: Potential for acidification of soil and water resources
EP: Eutrophication potential
POCP: Tropospheric ozone formation potential
ADPE: Abiotic resource depletion potential for non-fossil resources
ADFP: Abiotic resource depletion potential for fossil resources
PERT: Total use of renewable primary energy
PERNRT: Total use of non-renewable primary energy
FW: Net use of tap water resources

Table 4 Environmental impact and resource use indicators of the sub-base layer during the manufacturing process

Consumption		Life cycle stage									
		Manufacturing									
		A1. Raw materials supply A2. Transportation of raw materials A3. Product manufacturing									
		GWP	ODP	AP	EP	POCP	ADPE	ADFP	PERT	PERNRT	FW
		CO_2 eq. (kg)	CFC 11 eq. (kg)	SO_2 eq. (kg)	$(PO_4)^{3-}$ eq. (kg)	etileno eq. (kg)	Sb eq. (kg)	(MJ)	(MJ)	(MJ)	(m^3)
Materials	Weight (kg)										
Natural aggregate	2,200.000	15.400	2.20e−06	0.22	0.044	0.22	1.10e−05	264.00	15.40	264.00	0.660
Machinery	Volume (l)										
Total energy and emissions		15.400	2.20e−06	0.22	0.044	0.22	1.10e−05	264.00	15.40	264.00	0.660

Table 5 Environmental impact and resource use indicators of the sub-base layer during the construction process (transport)

Consumption		Life cycle stage									
		Construction									
		A4. Product transportation									
		GWP	ODP	AP	EP	POCP	ADPE	ADFP	PERT	PERNRT	FW
		CO_2 eq. (kg)	CFC 11 eq. (kg)	SO_2 eq. (kg)	$(PO_4)^{3-}$ eq. (kg)	etileno eq. (kg)	Sb eq. (kg)	(MJ)	(MJ)	(MJ)	(m^3)
Materials	Weight (kg)										
Natural aggregate	2,200.000	2.891	0.004	0.202	0.040	0.011	0.002	318.035		39.071	0.607
Machinery	Volume (l)										
Total energy and emissions		2.891	0.004	0.202	0.040	0.011	0.002	318.035		39.071	0.607

Table 6 Environmental impact and resource use indicators of the sub-base layer during the construction process (construction)

Consumption		Life cycle stage									
		Construction									
		A5. Construction and installation process									
		GWP	ODP	AP	EP	POCP	ADPE	ADFP	PERT	PERNRT	FW
		CO_2 eq. (kg)	CFC 11 eq. (kg)	SO_2 eq. (kg)	$(PO_4)^{3-}$ eq. (kg)	etileno eq. (kg)	Sb eq. (kg)	(MJ)	(MJ)	(MJ)	(m^3)
Materials	Weight (kg)										
Natural aggregate	2,200.000	0.016	1.04e−09	6.91e−05	2.83e−04	1.34e−05	5.81e−07	0.220		0.108	0.031
Machinery	Volume (l)										
Gasoil	0.968	2645	1.75e−07	0.012	0.048	0.002	9.79e−05	37.035		35.748	5.291
Auxiliary means		0.003	2.06e−10	1.37e-05	5.61e-05	2.65e−06	1.15e−07	0.044		0.021	0.006
Total energy and emissions		2.664	1.76e−07	0.012	0.048	0.002	9.86e−05	37.299		35.878	5.328

Regarding the environmental impact and resource utilization of the sub-base layer, detailed within are the specific embodied energy values of the material, along with their corresponding conversion factors (see Table 3). On one hand, the table illustrates environmental impact indicators related to global warming potential (GWP), stratospheric ozone depletion (ODP), soil and water resource acidification depletion (AP), eutrophication potential (EP), tropospheric ozone formation potential (POCP), abiotic resources for non-fossil resources depletion potential (ADPE), and abiotic resources for fossil resources depletion potential (ADFP). On the other hand, resource use indicators are provided concerning the total use of renewable primary energy (PERT), total use of non-renewable primary energy (PERNRT), and net use of freshwater resources (FW). All these indicators are itemised for various stages of the life cycle, classified as follows: raw material supply, raw material transportation and product manufacturing; product transportation; and construction and installation processes.

Subsequently, in Tables 4, 5, and 6, the environmental impact and resource use indicators of the sub-base layer are categorized by life cycle stage according to the materials, machinery, and auxiliary means utilized. In the manufacturing stage (Table 4) and the product transportation stage (Table 5), the values of these environmental impact and resource use indicators for the material, in this case natural aggregate, align with the total values shown in Table 1. However, in the construction and installation stage (Table 6), it can be observed that the fuel consumption of the machinery used is the primary source of energy and emissions in this stage of the sub-base layer's life cycle. The environmental impact values associated with auxiliary means closely resemble those generated by asphalt bitumen, indicating a relatively minor impact on the construction of the base layer compared to other components. The environmental impact values associated with auxiliary means are insignificant when juxtaposed with those stemming from natural aggregate and diesel, as their respective impact values are substantially smaller.

- Base layer

To continue the LCA process, the next step involves evaluating the impact of the base layer. Subsequently, in the broken-down prices tab of the menu, the material constituting the base layer to be analysed, hot mix bituminous AC, will be searched for. In this case, it can be found under the categories of urban pavements and pavements—urban pavements—bituminous mixtures and irrigation—hot mix bituminous AC. Once the material is identified, its properties will be configured, including the type and maximum size of aggregate (32 mm limestone), the type of layer (base layer) and the type of binder to use in the mixture (B35/50), as well as the type of mixture (semi-dense), the application tonnage (more than 1000 t/day) and the bitumen (0.05 t of bitumen per t of mixture). With the material parameters defined, its information, including breakdown price, specifications, generated waste, environmental impact indicators, and health and safety data, can be extracted. For this instance, data concerning the breakdown price, waste generated, and environmental impact indicators will solely be retrieved using the export option in PDF format situated on the right-hand side of the interface. Following that, the downloaded information will be analysed.

As in the sub-base layer, in the following table (Table 7), which identifies and quantifies the materials utilized for the base layer construction, a detailed description of the material of the respective layer is provided first, along with its identification code and unit of measurement. Additionally, in the second part of the table, the different materials comprising the layer are categorised, along with the equipment, machinery, and labour required for its execution, as well as the complementary direct costs stemming from it.

To assess waste generation, Table 8 provides detailed information on the waste generated from the execution of the layer, quantified in both mass (kg) and volume (l) units. As expected, since the base layer consists solely of bituminous mixtures, it constitutes the sole type of waste.

Regarding the environmental impact and resource utilization of the base layer, detailed within are the specific embodied energy values of the material, along with their corresponding conversion factors (see Table 9). Likewise, the table illustrates environmental impact and resource use indicators itemised for various stages of the life cycle.

Subsequently, in Tables 10, 11, and 12, the environmental impact and resource utilization metrics of the base layer are segmented by life cycle phase, categorized by the materials, machinery, and auxiliary resources employed. In the production phase (Table 10) and the transportation of the product phase (Table 11), the environmental impact and resource utilization metrics for the base layer are delineated based on the two constituent materials: asphalt bitumen and natural aggregate. Across most environmental impact indicators in these phases, asphalt bitumen exhibits notably higher values, except for the indicator pertaining to tropospheric ozone formation potential, where natural aggregate displays a greater impact. Regarding indicators concerning soil and water resource acidification depletion and eutrophication potential, the values for both materials are more evenly balanced. In terms of resource utilization indicators, there is greater variability: natural aggregate has a more pronounced impact on the indicator related to total renewable primary energy consumption, whereas asphalt bitumen demonstrates markedly higher values for the indicator concerning total non-renewable primary energy consumption. However, for the indicator of net utilization of freshwater resources, both materials exhibit similar impacts.

Table 7 Unit cost of the base layer

MPB001	t	Hot mix bituminous AC			

Continuous hot bituminous mix AC 32 base B35/50 S, for base layer, semi-dense, with limestone aggregate of 32 mm maximum size, with 0.05 t of bitumen per t of mixture, for an application tonnage of more than 1000 t/day. The price does not include transportation of the mixture

Code	Unit	Description	Performance	Unit price	Cost
1	*Materials*				
mt47aag001ogj	t	Continuous hot bituminous mix AC 32 base B35/50 S, for base layer, semi-dense, with limestone aggregate of 32 mm maximum size, with 0.05 t of bitumen per t of mixture, according to UNE-EN 13.108–1	1050	83.42	87.59
			Materials subtotal		87.59
2	*Equipment and machinery*				
mq11ext030	h	81 kW chain asphalt paver	0.011	231.73	2.55
mq02ron010a	h	Self-propelled tandem vibrating roller. 24.8 kW. 2450 kg. working width 100 cm	0.011	56.81	0.62
mq11com010	h	Self-propelled tire compactor, 12/22 t	0.011	66.47	0.73
			Equipment and machinery subtotal		3.90
3	*Labour*				
mo041	h	1st civil works construction officer	0.066	22.25	1.47
mo087	h	Civil works construction assistant	0.044	21.56	0.95
			Labour subtotal		2.42
4	Additional direct costs				
	%	Additional direct costs	2.000	93.91	1.88

(continued)

Table 7 (continued)

MPB001	t	Hot mix bituminous AC			

Continuous hot bituminous mix AC 32 base B35/50 S, for base layer, semi-dense, with limestone aggregate of 32 mm maximum size, with 0.05 t of bitumen per t of mixture, for an application tonnage of more than 1000 t/day. The price does not include transportation of the mixture

Code	Unit	Description	Performance	Unit price	Cost
Ten-year maintenance cost: 13.41€ in the first 10 years			Direct costs $(1 + 2 + 3 + 4)$		95.79
Reference and title of the standard			Applicability[a]	Obligation[b]	System[c]
EN 13.108–1:2006			1.3.2007	1.3.2008	
Bituminous mixtures. Material specifications. Part 1: Bituminous concrete					
EN 13.108-1:2006/AC:2008			1.1.2009	1.1.2009	1/2 + /3/4

[a]Date of applicability of the harmonized standard
[b]Date on which the coexistence period ends
[c]System for evaluating and verifying the constancy of benefits

Table 8 Waste generated by the base layer

LER code	Type	Weight (kg)	Volume (l)
17 03 02	Bituminous mix other than those mentioned in code 17 03 01	11.025	11.025
	Waste generated	11.025	11.025

However, during the construction and installation phase (Table 12), similar to what was observed in the previously analysed sub-base layer, it becomes evident that the fuel consumption of the utilized machinery stands out as the predominant source of energy consumption and emissions across all the indicators examined in this phase of the base layer's life cycle. Similarly, during this life cycle stage, the environmental impact of natural aggregate is significantly higher across all studied indicators, except for the abiotic resources for fossil resources depletion potential and the total use of non-renewable primary energy, where both materials exhibit more balanced values. The environmental impact values associated with auxiliary means closely resemble those generated by asphalt bitumen, indicating a relatively minor impact on the construction of the base layer compared to the other components.

Table 9 Environmental impact and resource use indicators of the base layer

Life cycle stage	Environmental impact indicators							Use of resources		
	GWP	ODP	AP	EP	POCP	ADPE	ADFP	PERT	PERNRT	FW
	CO_2 eq. (kg)	CFC 11 eq. (kg)	SO_2 eq. (kg)	$(PO_4)^{3-}$ eq. (kg)	etileno eq. (kg)	Sb eq. (kg)	(MJ)	(MJ)	(MJ)	(m^3)
Total A1-A2-A3	20.958	1.95e−06	0.205	0.022	0.083	9.26e−06	2,922.990	6.825	3,168.690	0.229
A4	3.450	0.004	0.241	0.048	0.013	0.002	379.474		46.618	0.724
A5	0.496	3.27e−08	0.002	0.009	4.21e−04	1.83e−05	6.939		6.637	0.991

(continued)

Table 9 (continued)

Life cycle stage	Environmental impact indicators							Use of resources		
	GWP	ODP	AP	EP	POCP	ADPE	ADFP	PERT	PERNRT	FW
	CO_2 eq. (kg)	CFC 11 eq. (kg)	SO_2 eq. (kg)	$(PO_4)^{3-}$ eq. (kg)	etileno eq. (kg)	Sb eq. (kg)	(MJ)	(MJ)	(MJ)	(m^3)
Total A4-A5	3.945	0.004	0.244	0.057	0.013	0.002	386.413		53.256	1.716
Total energy and emissions	24.903	0.004	0.449	0.080	0.096	0.002	3,309,403	6.825	3,221,946	1.945

A1: Raw materials supply
A2: Transportation of raw materials
A3: Product manufacturing
A4: Product transportation
A5: Construction and installation process
GWP: Global warming potential
ODP: Stratospheric ozone depletion potential
AP: Potential for acidification of soil and water resources
EP: Eutrophication potential
POCP: Tropospheric ozone formation potential
ADPE: Abiotic resource depletion potential for non-fossil resources
ADFP: Abiotic resource depletion potential for fossil resources
PERT:-Total use of renewable primary energy
PERNRT: Total use of non-renewable primary energy
FW: Net use of tap water resources

Table 10 Environmental impact and resource use indicators of the base layer during the manufacturing process

Consumption		Life cycle stage									
		Manufacturing									
		A1. Raw materials supply A2. Transportation of raw materials A3. Product manufacturing									
		GWP	ODP	AP	EP	POCP	ADPE	ADFP	PERT	PERNRT	FW
		CO_2 eq. (kg)	CFC 11 eq. (kg)	SO_2 eq. (kg)	$(PO_4)^{3-}$ eq. (kg)	etileno eq. (kg)	Sb eq. (kg)	(MJ)	(MJ)	(MJ)	(m^3)
Materials	Weight (kg)										
Asphalt bitumen	63.000	17.010	1.26e−06	0.126	0.013	0.004	6.30e−06	2,853.900	1.890	3,099.600	0.032
Natural aggregate	987.000	3.948	6.91e−07	0.079	0.010	0.079	2.96e−06	69.090	4.935	69.090	0.197
Total	1,050.000	20.958	1.95e−06	0.205	0.022	0.083	9.26e−06	2,922.990	6.825	3,168.690	0.229
Machinery	Volume (l)										
Total energy and emissions		20.958	1.95e−06	0.205	0.022	0.083	9.26e−06	2,922.990	6.825	3,168.690	0.229

Table 11 Environmental impact and resource use indicators of the base layer during the construction process (transport)

Consumption lePara>		Life cycle stage									
		Construction									
		A4. Product transportation									
		GWP	ODP	AP	EP	POCP	ADPE	ADFP	PERT	PERNRT	FW
		CO_2 eq. (kg)	CFC 11 eq. (kg)	SO_2 eq. (kg)	$(PO_4)^{3-}$ eq. (kg)	etileno eq. (kg)	Sb eq. (kg)	(MJ)	(MJ)	(MJ)	(m^3)
Materials	Weight (kg)										
Asphalt bitumen	63.000	0.207	2.69e−04	0.014	0.003	7.66e−04	1.24e−04	22.768		2.797	0.043
Natural aggregate	987.000	3.243	0.004	0.227	0.045	0.012	0.002	356.706		43.821	0.681
Total	1,050.000	3.450	0.004	0.241	0.048	0.013	0.002	379.474		46.618	0.724
Machinery	Volume (l)										
Total energy and emissions		3.450	0.004	0.241	0.048	0.013	0.002	379.474		46.618	0.724

Table 12 Environmental impact and resource use indicators of the base layer during the construction process (construction)

Consumption		Life cycle stage									
		Construction									
		A5. Construction and installation process									
		GWP	ODP	AP	EP	POCP	ADPE	ADFP	PERT	PERNRT	FW
		CO_2 eq. (kg)	CFC 11 eq. (kg)	SO_2 eq. (kg)	$(PO_4)^{3-}$ eq. (kg)	etileno eq. (kg)	Sb eq. (kg)	(MJ)	(MJ)	(MJ)	(m^3)
Materials	Weight (kg)										
Asphalt bitumen	63.000	4.50e−04	2.97e−11	1.98e−06	8.10e−06	3.82e−07	1.66e−08	0.006		0.003	9.00e−04
Natural aggregate	987.000	0.007	4.65e−10	3.10e−05	1.27e−04	5.99e−06	2.61e−07	0.099		0.049	0.014
Total	1,050.000	0.007	4.95e−10	3.30e−05	1.35e−04	6.37e−06	2.77e−07	0.105		0.052	0.015
Machinery	Volume (l)										
Gasoil	0.178	0.486	3.21e−08	0.002	0.009	4.13e−04	1.80e−05	6.810		6.574	0.973
Auxiliary means		0.002	1.14e−10	7.58e−06	3.10e−05	1.46e−06	6.37e−08	0.024		0.012	0.003
Total energy and emissions		0.496	3.27e−08	0.002	0.009	4.21e−04	1.83e−05	6.939		6.637	0.991

- Intermediate layer

To continue the LCA process, the next step involves evaluating the impact of the intermediate layer. Subsequently, in the broken-down prices tab of the menu, the material constituting the intermediate layer to be analysed, hot mix bituminous AC, will be searched for. In this case, it can be found at the same place as the base layer, under the categories of urban pavements and pavements—urban pavements—bituminous mixtures and irrigation—hot mix bituminous AC. Once the material is identified, its properties will be configured, including the type and maximum size of aggregate (32 mm limestone), the type of layer (intermediate layer) and the type of binder to use in the mixture (B35/50), as well as the type of mixture (semi-dense), the application tonnage (more than 1000 t/day) and the bitumen (0.05 t of bitumen per t of mixture). With the material parameters defined, its information, including breakdown price, specifications, generated waste, environmental impact indicators, and health and safety data, can be extracted. For this instance, as in the previous layers, data concerning the breakdown price, waste generated, and environmental impact indicators will solely be retrieved using the export option in PDF format situated on the right-hand side of the interface. Following that, the downloaded information will be analysed.

As in the previous layers, in the following table (Table 13), which identifies and quantifies the materials utilized for the intermediate layer construction, a detailed description of the material of the respective layer is provided first, along with its identification code and unit of measurement. Additionally, in the second part of the table, the different materials comprising the layer are categorised, along with the equipment, machinery, and labour required for its execution, as well as the complementary direct costs stemming from it.

To assess waste generation, Table 14 provides detailed information on the waste generated from the execution of the layer, quantified in both mass (kg) and volume (l) units. As expected, since the base layer consists solely of bituminous mixtures, it constitutes the sole type of waste.

Regarding the environmental impact and resource utilization of the intermediate layer, detailed within are the specific embodied energy values of the material, along with their corresponding conversion factors (see Table 15). Likewise, the table illustrates environmental impact and resource use indicators itemised for various stages of the life cycle.

Subsequently, in Tables 16, 17, and 18, the environmental impact and resource consumption metrics of the intermediate layer are organized by life cycle phase, characterized by the materials, machinery, and auxiliary resources utilized. During the production phase (Table 16) and the transportation of the product phase (Table 17), the

Table 13 Unit cost of the intermediate layer

MPB001	t	Hot mix bituminous AC			

Continuous hot bituminous mix AC 32 bin B35/50 S, for intermediate layer, semi-dense, with limestone aggregate of 32 mm maximum size, with 0.05 t of bitumen per t of mixture, for an application tonnage of more than 1000 t/day. The price does not include transportation of the mixture

Code	Unit	Description	Performance	Unit price	Cost
1	*Materials*				
mt47aag001ngj	t	Continuous hot bituminous mix AC 32 bin B35/50 S, for intermediate layer, semi-dense, with limestone aggregate of 32 mm maximum size, with 0.05 t of bitumen per t of mixture, according to UNE-EN 13,108–1	1.050	83.42	87.59
			Materials subtotal:	87.59	
2	*Equipment and machinery*				
mq11ext030	h	81 kW chain asphalt paver	0.011	231.73	2.55
mq02ron010a	h	Self-propelled tandem vibrating roller, 24.8 kW, 2450 kg, working width 100 cm	0.011	56.81	0.62
mq11com010	h	Self-propelled tire compactor, 12/22 t	0.011	66.47	0.73
			Equipment and machinery subtotal:	3.90	
3	*Labour*				
mo041	h	1st civil works construction officer	0.066	22.25	1.47
mo087	h	Civil works construction assistant	0.044	21.56	0.95
			Labour subtotal:	2.42	
4		Additional direct costs			

(continued)

Table 13   (continued)

MPB001	t	Hot mix bituminous AC			

Continuous hot bituminous mix AC 32 bin B35/50 S, for intermediate layer, semi-dense, with limestone aggregate of 32 mm maximum size, with 0.05 t of bitumen per t of mixture, for an application tonnage of more than 1000 t/day. The price does not include transportation of the mixture

Code	Unit	Description	Performance	Unit price	Cost
	%	Additional direct costs	2000	93.91	1.88
Ten-year maintenance cost: 13.41€ in the first 10 years			Direct costs (1 + 2 + 3 + 4):		95.79
Reference and title of the standard			Applicability[a]	Obligation[b]	System[c]
EN 13,108–1:2006			1.3.2007	1.3.2008	
Bituminous mixtures. Material specifications. Part 1: Bituminous concrete					
EN 13,108–1:2006/AC:2008			1.1.2009	1.1.2009	1/2+/3/4

[a]Date of applicability of the harmonized standard
[b]Date on which the coexistence period ends
[c]System for evaluating and verifying the constancy of benefits

Table 14   Waste generated by the intermediate layer

LER code	Type	Weight (kg)	Volume (l)
17 03 02	Bituminous mix other than those mentioned in code 17 03 01	11.025	11.025
	Waste generated	11.025	11.025

environmental impact and resource consumption metrics for the intermediate layer are delineated based on the two constituent materials: asphalt bitumen and natural aggregate. Unlike the sub-base and base layers, in the manufacturing phase of this intermediate layer, asphalt bitumen exhibits higher impact values than natural aggregates, whereas during the construction phase, natural aggregate demonstrates higher impact values across all environmental impact and resource consumption indicators.During the production phase (Table 16), asphalt bitumen generally shows significantly higher values across most environmental impact indicators compared to natural aggregate. However, there are exceptions where both materials exhibit more balanced values, particularly in indicators related to soil and water resource acidification depletion, eutrophication potential, tropospheric ozone formation potential, and abiotic resources for non-fossil resources depletion potential. Resource utilization indicators show greater variability: natural aggregate tends to have a more pronounced impact on the indicator concerning total renewable primary energy consumption, while asphalt bitumen typically demonstrates substantially higher values for the indicator of total non-renewable primary energy consumption. Nonetheless, for the

P. Porras-Pereira and P. Mercader-Moyano

Table 15 Environmental impact and resource use indicators of the intermediate layer

Life cycle stage	Environmental impact indicators							Use of resources		
	GWP	ODP	AP	EP	POCP	ADPE	ADFP	PERT	PERNRT	FW
	CO_2 eq. (kg)	CFC 11 eq. (kg)	SO_2 eq. (kg)	$(PO_4)^{3-}$ eq. (kg)	etileno eq. (kg)	Sb eq. (kg)	(MJ)	(MJ)	(MJ)	(m^3)
Total A1-A2-A3	20.958	1.95e−06	0.205	0.022	0.083	9.26e-06	2,922.990	6.825	3,168.690	0.229
A4	3.450	0.004	0.241	0.048	0.013	0.002	379.474		46.618	0.724
A5	0.496	3.27e−08	0.002	0.009	4.21e−04	1.83e−05	6.939		6.637	0.991

(continued)

Table 15 (continued)

Life cycle stage	Environmental impact indicators							Use of resources		
	GWP	ODP	AP	EP	POCP	ADPE	ADFP	PERT	PERNRT	FW
	CO_2 eq. (kg)	CFC 11 eq. (kg)	SO_2 eq. (kg)	$(PO_4)^{3-}$ eq. (kg)	etileno eq. (kg)	Sb eq. (kg)	(MJ)	(MJ)	(MJ)	(m^3)
Total A4-A5	3.945	0.004	0.244	0.057	0.013	0.002	386.413		53.256	1.716
Total energy and emissions	24.903	0.004	0.449	0.080	0.096	0.002	3,309.403	6.825	3,221.946	1.945

A1: Raw materials supply
A2: Transportation of raw materials
A3: Product manufacturing
A4: Product transportation
A5: Construction and installation process
GWP: Global warming potential
ODP: Stratospheric ozone depletion potential
AP: Potential for acidification of soil and water resources
EP: Eutrophication potential
POCP: Tropospheric ozone formation potential
ADPE: Abiotic resource depletion potential for non-fossil resources
ADFP: Abiotic resource depletion potential for fossil resources
PERT: Total use of renewable primary energy
PERNRT: Total use of non-renewable primary energy
FW: Net use of tap water resources

Table 16 Environmental impact and resource use indicators of the intermediate layer during the manufacturing process

Consumption		Life cycle stage									
		Manufacturing									
		A1. Raw materials supply A2. Transportation of raw materials A3. Product manufacturing									
		GWP	ODP	AP	EP	POCP	ADPE	ADFP	PERT	PERNRT	FW
		CO_2 eq. (kg)	CFC 11 eq. (kg)	SO_2 eq. (kg)	$(PO_4)^{3-}$ eq. (kg)	etileno eq. (kg)	Sb eq. (kg)	(MJ)	(MJ)	(MJ)	(m^3)
Materials	Weight (kg)										
Asphalt bitumen	63.000	17.010	1.26e−06	0.126	0.013	0.004	6.30e−06	2,853.900	1.890	3,099.600	0.032
Natural aggregate	987.000	3.948	6.91e−07	0.079	0.010	0.079	2.96e−06	69.090	4.935	69.090	0.197
Total	1,050.000	20.958	1.95e−06	0.205	0.022	0.083	9.26e−06	2,922.990	6.825	3,168.690	0.229
Machinery	Volume (l)										
Total energy and emissions		20.958	1.95e−06	0.205	0.022	0.083	9.26e−06	2,922.990	6.825	3,168.690	0.229

Table 17 Environmental impact and resource use indicators of the intermediate layer during the construction process (transport)

Consumption		Life cycle stage									
		Construction									
		A4. Product transportation									
		GWP	ODP	AP	EP	POCP	ADPE	ADFP	PERT	PERNRT	FW
		CO_2 eq. (kg)	CFC 11 eq. (kg)	SO_2 eq. (kg)	$(PO_4)^{3-}$ eq. (kg)	etileno eq. (kg)	Sb eq. (kg)	(MJ)	(MJ)	(MJ)	(m^3)
Materials	Weight (kg)										
Asphalt bitumen	63.000	0.207	2.69e−04	0.014	0.003	7.66e−04	1.24e−04	22.768		2.797	0.043
Natural aggregate	987.000	3.243	0.004	0.227	0.045	0.012	0.002	356.706		43.821	0.681
Total	1,050.000	3.450	0.004	0.241	0.048	0.013	0.002	379.474		46.618	0.724
Machinery	Volume (l)										
Total energy and emissions		3.450	0.004	0.241	0.048	0.013	0.002	379.474		46.618	0.724

Table 18 Environmental impact and resource use indicators of the intermediate layer during the construction process (construction)

Consumption		Life cycle stage									
		Construction									
		A5. Construction and installation process									
		GWP	ODP	AP	EP	POCP	ADPE	ADFP	PERT	PERNRT	FW
		CO_2 eq. (kg)	CFC 11 eq. (kg)	SO_2 eq. (kg)	$(PO_4)^{3-}$ eq. (kg)	etileno eq. (kg)	Sb eq. (kg)	(MJ)	(MJ)	(MJ)	(m^3)
Materials	Weight (kg)										
Asphalt bitumen	63.000	4.50e−04	2.97e−11	1.98e−06	8.10e−06	3.82e−07	1.66e−08	0.006		0.003	9.00e−04
Natural aggregate	987.000	0.007	4.65e−10	3.10e−05	1.27e−04	5.99e−06	2.61e−07	0.099		0.049	0.014
Total	1,050.000	0.007	4.95e−10	3.30e−05	1.35e−04	6.37e−06	2.77e−07	0.105		0.052	0.015
Machinery	Volume (l)										
Gasoil	0.178	0.486	3.21e−08	0.002	0.009	4.13e−04	1.80e−05	6.810		6.574	0.973
Auxiliary means		0.002	1.14e−10	7.58e−06	3.10e−05	1.46e−06	6.37e−08	0.024		0.012	0.003
Total energy and emissions		0.496	3.27e−08	0.002	0.009	4.21e−04	1.83e−05	6.939		6.637	0.991

indicator measuring the net use of freshwater resources, both materials demonstrate similar impacts.

Conversely, during the construction phase (Table 17), natural aggregate displays significantly higher values than asphalt bitumen across most environmental impact indicators. However, there are exceptions where both materials demonstrate more balanced values, particularly in indicators related to eutrophication potential and abiotic resources for non-fossil resources depletion potential. Similarly, in terms of resource utilization indicators, natural aggregate consistently exhibits notably higher values than asphalt bitumen.

However, in the construction and installation phase (Table 18), mirroring observations from previous layers analysed, it becomes apparent that the fuel consumption of the utilized machinery emerges as the primary driver of energy consumption and emissions across all the examined indicators in this phase of the intermediate layer's life cycle. Likewise, during this life cycle stage, the environmental impact of natural aggregate is notably higher across all studied indicators, with exceptions noted in the abiotic resources for fossil resources depletion potential and the total use of non-renewable primary energy, where both materials exhibit more balanced values. The impact indicator values associated with auxiliary means closely resemble those generated by asphalt bitumen, highlighting their significance in the construction of the intermediate layer.

- Surfacing layer

In order to conclude the LCA process, the last step involves evaluating the impact of the surfacing layer. Subsequently, in the broken-down prices tab of the menu, the material constituting the surfacing layer to be analysed, hot mix bituminous AC, will be searched for. In this case, it can be found at the same place as the base and intermediate layer, under the categories of urban pavements and pavements—urban pavements—bituminous mixtures and irrigation—hot mix bituminous AC. Once the material is identified, its properties will be configured, including the type and maximum size of aggregate (22 mm limestone), the type of layer (surfacing layer) and the type of binder to use in the mixture (B35/50), as well as the type of mixture (semi-dense), the application tonnage (more than 1000 t/day) and the bitumen (0.05 t of bitumen per t of mixture). With the material parameters defined, its information, including breakdown price, specifications, generated waste, environmental impact indicators, and health and safety data, can be extracted. For this instance, as in the previous layers, data concerning the breakdown price, waste generated, and environmental impact indicators will solely be retrieved using the export option in PDF format situated on the right-hand side of the interface. Following that, the downloaded information will be analysed.

As in the previous layers, in the following table (Table 19), which identifies and quantifies the materials utilized for the surfacing layer construction, a detailed description of the material of the respective layer is provided first, along with its identification code and unit of measurement. Additionally, in the second part of the table, the different materials comprising the layer are categorised, along with the equipment, machinery, and labour required for its execution, as well as the complementary direct costs stemming from it.

To assess waste generation, Table 20 provides detailed information on the waste generated from the execution of the layer, quantified in both mass (kg) and volume (l) units. As expected, since the surfacing layer consists solely of bituminous mixtures, it constitutes the sole type of waste.

Regarding the environmental impact and resource utilization of the surfacing layer. detailed within are the specific embodied energy values of the material. along with their corresponding conversion factors (see Table 21). Likewise. the table illustrates environmental impact and resource use indicators itemised for various stages of the life cycle.

Subsequently, in Tables 22, 23 and 24, the environmental impact and resource utilization metrics of the surfacing layer are presented by life cycle phase, reflecting the materials, machinery, and auxiliary resources involved. During the production phase (Table 22) and the transportation of the product phase (Table 23), the environmental impact and resource utilization metrics for the surfacing layer are detailed, focusing on the two constituent materials: asphalt bitumen and natural aggregate. Similar to the intermediate layer, during the manufacturing phase of this surfacing layer, asphalt bitumen demonstrates higher impact values than natural aggregates, whereas in the construction phase, natural aggregate exhibits higher impact values across all environmental impact and resource utilization indicators.

During the production phase (Table 22), asphalt bitumen generally exhibits notably higher values across most environmental impact indicators compared to natural aggregate. Nevertheless, there are instances where both materials display more balanced values, particularly in indicators related to soil and water resource acidification depletion, eutrophication potential, tropospheric ozone formation potential, and abiotic resource depletion potential for non-fossil resources. Moreover, resource utilization indicators demonstrate greater variability: natural aggregate tends to exert a more significant impact on the indicator concerning total renewable primary energy consumption, while asphalt bitumen typically presents substantially higher values for the indicator of total non-renewable primary energy consumption. However, for the indicator measuring the net use of freshwater resources, both materials exhibit similar impacts.

Contrarywise, in the construction phase (as illustrated in Table 23), natural aggregate exhibits notably higher values than asphalt bitumen across most environmental impact indicators. Nevertheless, there are instances where both materials present more balanced values, particularly in indicators related to soil and water resource acidification depletion and eutrophication potential. Similarly, regarding resource utilization indicators, natural aggregate consistently demonstrates notably higher values than asphalt bitumen.

However, during the construction and installation phase (as depicted in Table 24), resembling findings from previous layers analysed, it becomes evident that the fuel consumption of the utilized machinery emerges as the primary contributor to energy consumption and emissions across all the examined indicators in this phase of the

Table 19 Unit cost of the surfacing layer

MPB001	t	Hot mix bituminous AC			

Continuous hot bituminous mix AC 22 surf B35/50 S, for surfacing layer, semi-dense, with limestone aggregate of 22 mm maximum size, with 0.05 t of bitumen per t of mixture, for an application tonnage of more than 1000 t/day. The price does not include transportation of the mixture

Code	Unit	Description	Performance	Unit price	Cost
1	*Materials*				
mt47aag001jgj	t	Continuous hot bituminous mix AC 22 surf B35/50 S, for surfacing layer, semi-dense, with limestone aggregate of 22 mm maximum size, with 0.05 t of bitumen per t of mixture, according to UNE-EN 13.108–1	1.050	83.47	87.64
			Materials subtotal	87.64	
2	*Equipment and machinery*				
mq11ext030	h	81 kW chain asphalt paver	0.011	231.73	2.55
mq02ron010a	h	Self-propelled tandem vibrating roller. 24.8 kW, 2450 kg, working width 100 cm	0.011	56.81	0.62
mq11com010	h	Self-propelled tire compactor, 12/22 t	0.011	66.47	0.73
			Equipment and machinery subtotal	3.90	
3	*Labour*				
mo041	h	1st civil works construction officer	0.066	22.25	1.47
mo087	h	Civil works construction assistant	0.044	21.56	0.95
			Labour subtotal	2.42	
4	*Additional direct costs*				

(continued)

Table 19 (continued)

MPB001	t	Hot mix bituminous AC			

Continuous hot bituminous mix AC 22 surf B35/50 S, for surfacing layer, semi-dense, with limestone aggregate of 22 mm maximum size, with 0.05 t of bitumen per t of mixture, for an application tonnage of more than 1000 t/day. The price does not include transportation of the mixture

Code	Unit	Description	Performance	Unit price	Cost
	%	Additional direct costs	2.000	93.96	1.88
Ten-year maintenance cost: 13.42€ in the first 10 years			Direct costs (1 + 2 + 3 + 4)		95.84
Reference and title of the standard			Applicability[a]	Obligation[b]	System[c]
EN 13.108–1:2006			1.3.2007	1.3.2008	
Bituminous mixtures. Material specifications. Part 1: Bituminous concrete					
EN 13.108–1:2006/AC:2008			1.1.2009	1.1.2009	1/2+/3/4

[a]Date of applicability of the harmonized standard
[b]Date on which the coexistence period ends
[c]System for evaluating and verifying the constancy of benefits

Table 20 Waste generated by the surfacing layer

LER code	Type	Weight (kg)	Volume (l)
17. 03.02	Bituminous mix other than those mentioned in code 17 03 01	11.025	11.025
	Waste generated	11.025	11.025

surfacing layer's life cycle. Similarly, during this life cycle stage, the environmental impact of natural aggregate is notably higher across all studied indicators, with the exception noted in the total use of non-renewable primary energy, where both materials exhibit more balanced values. The impact indicator values associated with auxiliary means closely resemble those generated by asphalt bitumen, underscoring their significance in the construction of the surfacing layer.

After completing the LCA for an unmodified asphalt mixture, it becomes evident that the values for each environmental impact indicator and resource usage across the base, intermediate, and surfacing layers remain consistent. This uniformity arises from the identical composition of these top three layers (asphalt bitumen and natural aggregate), which only modify their granulometry, resulting in equivalent environmental impact and resource consumption profiles.

Table 21 Environmental impact and resource use indicators of the surfacing layer

Life cycle stage	Environmental impact indicators							Use of resources		
	GWP	ODP	AP	EP	POCP	ADPE	ADFP	PERT	PERNRT	FW
	CO_2 eq. (kg)	CFC 11 eq. (kg)	SO_2 eq. (kg)	$(PO_4)^{3-}$ eq. (kg)	etileno eq. (kg)	Sb eq. (kg)	(MJ)	(MJ)	(MJ)	(m^3)
Total A1-A2-A3	20.958	1.95e−06	0.205	0.022	0.083	9.26e−06	2,922.990	6.825	3,168.690	0.229
A4	3.450	0.004	0.241	0.048	0.013	0.002	379.474		46.618	0.724
A5	0.496	3.27e−08	0.002	0.009	4.21e−04	1.83e−05	6.939		6.637	0.991

(continued)

P. Porras-Pereira and P. Mercader-Moyano

Table 21 (continued)

Life cycle stage	Environmental impact indicators							Use of resources		
	GWP	ODP	AP	EP	POCP	ADPE	ADFP	PERT	PERNRT	FW
	CO_2 eq. (kg)	CFC 11 eq. (kg)	SO_2 eq. (kg)	$(PO_4)^{3-}$ eq. (kg)	etileno eq. (kg)	Sb eq. (kg)	(MJ)	(MJ)	(MJ)	(m^3)
Total A4-A5	3.945	0.004	0.244	0.057	0.013	0.002	386.413		53.256	1.716
Total energy and emissions	24.903	0.004	0.449	0.080	0.096	0.002	3,309.403	6.825	3,221.946	1.945

A1: Raw materials supply
A2: Transportation of raw materials
A3: Product manufacturing
A4: Product transportation
A5: Construction and installation process
GWP: Global warming potential
ODP: Stratospheric ozone depletion potential
AP: Potential for acidification of soil and water resources
EP: Eutrophication potential
POCP: Tropospheric ozone formation potential
ADPE: Abiotic resource depletion potential for non-fossil resources
ADFP: Abiotic resource depletion potential for fossil resources
PERT: Total use of renewable primary energy
PERNRT: Total use of non-renewable primary energy
FW: Net use of tap water resources

Table 22 Environmental impact and resource use indicators of the surfacing layer during the manufacturing process

Consumption		Life cycle stage									
		Manufacturing									
		A1. Raw materials supply A2. Transportation of raw materials A3. Product manufacturing									
		GWP	ODP	AP	EP	POCP	ADPE	ADFP	PERT	PERNRT	FW
		CO_2 eq. (kg)	CFC 11 eq. (kg)	SO_2 eq. (kg)	$(PO_4)^{3-}$ eq. (kg)	etileno eq. (kg)	Sb eq. (kg)	(MJ)	(MJ)	(MJ)	(m^3)
Materials	Weight (kg)										
Asphalt bitumen	63.000	17.01	1.26e−06	0.126	0.013	0.004	6.30e−06	2,853.900	1.89	3,099.600	0.032
Natural aggregate	987.000	3.948	6.91e−07	0.079	0.01	0.079	2.96e−06	69.09	4.935	69.09	0.197
Total	1,050.000	20.958	1.95e−06	0.205	0.022	0.083	9.26e−06	2,922.990	6.825	3,168.690	0.229
Machinery	Volume (l)										
Total energy and emissions		20.958	1.95E−06	0.205	0.022	0.083	9.26E−06	2,922.990	6.825	3,168.690	0.229

Table 23 Environmental impact and resource use indicators of the surfacing layer during the construction process (transport)

Consumption		Life cycle stage									
		Construction									
		A4. Product transportation									
		GWP	ODP	AP	EP	POCP	ADPE	ADFP	PERT	PERNRT	FW
		CO_2 eq. (kg)	CFC 11 eq. (kg)	SO_2 eq. (kg)	$(PO_4)^{3-}$ eq. (kg)	etileno eq. (kg)	Sb eq. (kg)	(MJ)	(MJ)	(MJ)	(m^3)
Materials	Weight (kg)										
Asphalt bitumen	63.000	0.207	2.69e−04	0.014	0.003	7.66e−04	1.24e−04	22.768		2.797	0.043
Natural aggregate	987.000	3.243	0.004	0.227	0.045	0.012	0.002	356.706		43.821	0.681
Total	1,050.000	3.450	0.004	0.241	0.048	0.013	0.002	379.474		46.618	0.724
Machinery	Volume (l)										
Total energy and emissions		3.450	0.004	0.241	0.048	0.013	0.002	379.474		46.618	0.724

Table 24 Environmental impact and resource use indicators of the surfacing layer during the construction process (construction)

Consumption		Life cycle stage									
		Construction									
		A5. Proceso de construcción e instalación									
		GWP	ODP	AP	EP	POCP	ADPE	ADFP	PERT	PERNRT	FW
		CO_2 eq. (kg)	CFC 11 eq. (kg)	SO_2 eq. (kg)	$(PO_4)^{3-}$ eq. (kg)	etileno eq. (kg)	Sb eq. (kg)	(MJ)	(MJ)	(MJ)	(m^3)
Materials	Weight (kg)										
Asphalt bitumen	63.000	4.50e−04	2.97e−11	1.98e−06	8.10e−06	3.82e−07	1.66e−08	0.006		0.003	9.00e−04
Natural aggregate	987.000	0.007	4.65e−10	3.10e−05	1.27e−04	5.99e−06	2.61e−07	0.099		0.049	0.014
Total	1,050,000	0.007	4.95e−10	3.30e−05	1.35e−04	6.37e−06	2.77e−07	0.105		0.052	0.015
Machinery	Volume (l)										
Gasoil	0.178	0.486	3.21e−08	0.002	0.009	4.13e−04	1.80e−05	6.810		6574	0.973
Auxiliary means		0.002	1.14e−10	7.58e−06	3.10e−05	1.46e−06	6.37e−08	0.024		0.012	0.003
Total energy and emissions		0.496	3.27e−08	0.002	0.009	4.21e−04	1.83e−05	6.939		6.637	0.991

2 Life Cycle Assessment of a Nano-Silica-Modified Asphalt Mixture

Upon completion of the LCA for an unmodified asphalt mixture, confirming the substantial environmental and resource utilization impact of asphalt bitumen throughout different stages of its life cycle, especially in the initial phases, the subsequent course of action entails conducting a comparable evaluation for a nano-silica-modified asphalt mixture. Due to the absence of materials with nanoparticles in the 'Construction Price Generator' software, data concerning the LCA of a nano-silica-modified asphalt mixture will be sourced from published scientific literature. However, it's worth noting that this data could have been sourced from any database containing such information. By adopting this approach, it becomes possible to ascertain the asphalt variant that is most environmentally preferable. This methodology proves to be applicable across diverse geographical regions and databases, provided that requisite data is accessible.

The impact values for each of the various categories employed in the LCA of a nano-silica-modified asphalt mixture will be sourced from the scientific article entitled "Life Cycle Assessment for the Production Phase of Nano-Silica-Modified Asphalt Mixtures," authored by Sackey Solomon, Lee Dong-Eun, and Kim Byung-Soo, who are associated with Kyungpook National University of Korea. This article was published in the MDPI "Applied Sciences" Journal on March 29th, 2019. For this study, as well as for the previous LCA, a Functional Unit of 1000 kg (1 t) of asphalt mixtures production was adopted.

Table 25 Environmental Impact Categories [1]

Impact Category	Reference Unit	Impact Result
Environmental impact|global warming	kg CO2-eq	7.44563×10^3
Human health|respiratory effects, average	kg PM2.5-Eq	8.86935×10^2
Environmental impact|ozone depletion	kg CFC-11-Eq	3.71600×10^{-2}
Environmental impact|eutrophication	kg N-Eq	1.49156×10^1
Human health|carcinogenic	kg benzene-Eq	2.18467×10^3
Environmental impact|photochemical oxidation	kg NOx-Eq	3.03420×10^1
Human health|non-carcinogenics	kg toluene-Eq	6.07040×10^6
Environmental impact|ecotoxicity	kg 2.4-D-Eq	1.08917×10^4
Environmental impact|acidification	moles of H + -Eq	1.87879×10^5

The forthcoming table (Table 25) displays the different environmental impact indicators published in the referenced scientific article, along with their respective impact values. These values will serve as the basis for comparison with the outcomes derived from the LCA of an unmodified asphalt mixture.

In the ensuing table (Table 26), adopting a similar format to that employed in the LCA of an unmodified asphalt mixture, a condensed summary of Table 25 is provided, exclusively presenting the values of environmental impact indicators shared by both LCAs. These indicators will be pivotal in facilitating a comparative analysis, aiding in the determination of the asphalt variant with the highest environmental preference.

As depicted in the preceding table (Table 26), this scientific publication does not provide data categorized by each of the life cycle stage, but instead presents overall values for the product's entire lifespan. Therefore, the comparison between both case studies will be conducted using the total values for each asphalt mixture.

Table 26 Environmental impact and resource use indicators of the nano-silica-modified asphalt mixture

Life cycle stage	Environmental impact indicators			
	GWP	ODP	AP	EP
	CO_2 eq. (kg)	CFC 11 eq. (kg)	SO_2 eq. (kg)	$(PO_4)^{3-}$ eq. (kg)
Total A1-A2-A3	–	–	–	–
A4	–	–	–	–
A5	–	–	–	–
Total A4-A5	–	–	–	–
Total energy and emissions	744.563×10^3	371.600×10^2	187.879×10^5	149.156×10^1

A1: Raw materials supply
A2: Transportation of raw materials
A3: Product manufacturing
A4: Product transportation
A5: Construction and installation process
GWP: Global warming potential
ODP: Stratospheric ozone depletion potential
AP: Potential for acidification of soil and water resources
EP: Eutrophication potentia

3 Quantitative Assessment of the Attained Outcomes

Subsequent to the completion of the LCA of an unmodified asphalt mixture and the acquisition of results from a LCA of a nano-silica-modified asphalt mixture, a graphical representation is employed to compare the results obtained for each case study across their entire life cycle, in order to discern the most ecologically sound option.

As indicated in the preceding section, due to the absence of categorized data for each life cycle stage in the consulted scientific publication, the comparison between both case studies will rely on the total values for each of the asphalt mixtures compared. Consequently, the results obtained from the LCA of an unmodified asphalt mixture for each layer of the road asphalt process (unit of work) are consolidated into a single table (Table 27). Due to the absence of asphalt mixture and the exclusive use of artificial granitic gravel in the sub-base layer, the results obtained for this layer will not undergo evaluation.

Finally, a graphical comparison is generated to illustrate the results obtained for each case study throughout their entire life cycle, focusing on the impact categories shared by both LCAs: global warming potential, stratospheric ozone depletion potential, potential for acidification of soil and water resources, and eutrophication potential (Table 28).

After evaluating a conventional asphalt mixture in terms of materials production emissions through the LCA methodology and comparing these findings with those of a nano-silica-modified asphalt mixture, the research underscores significant disparities in their impact contributions. Notably, the nano-silica-modified variant exhibits notably worse performance across all four analysed impact categories, particularly in the categories of global warming potential and stratospheric ozone depletion potential.

The category demonstrating the most pronounced disparity lies in the potential for acidification of soil and water resources, where the nano-silica-modified asphalt mixture registers an impact magnitude exceeding conventional asphalt by a factor of 100.000. Similarly, the analysis reveals substantial variability in the impact category of stratospheric ozone depletion potential, on account of the near-negligible impact of unmodified asphalt, which is almost zero. Furthermore, notable disparities are also observed in the category of global warming potential, where the impact of the nano-silica-modified asphalt mixture is 10 times greater. Conversely, the eutrophication potential category showcases more balanced values in both asphalt mixtures.

However, the conclusions derived from this research concerning nano-silica should not be extrapolated to all other types of nanomaterials. The environmental effects and overall outcomes associated with the integration of nanomaterials into asphalt depend on the manufacturing methods specific to each nanomaterial type.

Table 27 Environmental impact and resource use indicators of the base, intermediate and surfacing layers

Life cycle stage	Environmental impact indicators							Use of resources		
	GWP	ODP	AP	EP	POCP	ADPE	ADFP	PERT	PERNRT	FW
	CO_2 eq. (kg)	CFC 11 eq. (kg)	SO_2 eq. (kg)	$(PO_4)^{3-}$ eq. (kg)	etileno eq. (kg)	Sb eq. (kg)	(MJ)	(MJ)	(MJ)	(m^3)
Total A1-A2-A3	62.874	5.85e−06	0.615	0.066	0.249	2.78e−05	8,768.97	20.475	9,506.07	0.687
A4	1.035	0.012	0.723	0.144	0.039	0.006	1,138.422		139.854	2.172
A5	1.488	9.81e−08	0.006	0.027	1.26e−03	5.49e−05	20.817		19.911	2.973
Total A4-A5	11.835	0.012	0.732	0.171	0.039	0.006	1,159.239		159.768	5.148
Total energy and emissions	74.709	0.012	1.347	0.24	0.288	0.006	9,928.209	20.475	9,665.838	5.835

A1: Raw materials supply
A2: Transportation of raw materials
A3: Product manufacturing
A4: Product transportation
A5: Construction and installation process
GWP: Global warming potential
ODP: Stratospheric ozone depletion potential
AP: Potential for acidification of soil and water resources
EP: Eutrophication potential
POCP: Tropospheric ozone formation potential
ADPE: Abiotic resource depletion potential for non-fossil resources
ADFP: Abiotic resource depletion potential for fossil resources
PERT: Total use of renewable primary energy
PERNRT: Total use of non-renewable primary energy
FW: Net use of tap water resources

Table 28 Comparison of the environmental impact and resource use indicators of the upper layers

	Environmental impact indicators			
	GWP (Global warming potential)	ODP (Stratospheric ozone depletion potential)	AP (Potential for acidification of soil and water resources)	EP (Eutrophication potential)
	CO_2 eq. (kg)	CFC 11 eq. (kg)	SO_2 eq. (kg)	$(PO_4)^{3-}$ eq. (kg)
Unmodified asphalt mixture	74.709	0.012	1.347	0.24
Nano-silica-modified asphalt mixture	7,445.63	371.60	187,879.00	14.92

Reference

1. S. Solomon, L. Dong-Eun, K. Byung-Soo, Life Cycle Assessment for the Production Phase of Nano-Silica-Modified Asphalt Mixtures. Appl. Sci. **9**(7), 1315 (2019)

Conclusions

Building upon the earlier progress made in Life Cycle Analysis focusing on nanoparticles applied to the construction industry, the research has yielded detailed and precise information. This information is intended to be instrumental in educating and training professionals within the construction sector through a holistic approach, particularly focusing on environmental health and safety, thus aligning with the primary objective of the research.

Furthermore, these insights not only contribute to educate and train professionals, but also serve as a foundation for enhancing industry practices in the construction sector. The comprehensive approach to environmental health and safety, central to the research objective, aims to foster a sustainable and responsible integration of nanoparticles in construction materials and practices.

As aforementioned, LCA is a tool that helps to assess the environmental impacts of materials and products so that decisions can be made not just on the benefits of using these materials but also considering their environmental contributions. This study assessed a conventional asphalt mixture in terms of materials production emissions through the LCA methodology, and the results were compared to a nano-silica-modified asphalt mixtures to understand the impact contribution of nano-silica-modified asphalt mixtures. The results showed that the modification of asphalt materials with nano-silica causes the nano-silica-modified asphalt mixture exhibits notably worse performance across all four analysed impact categories. Nevertheless, the conclusions drawn from this research concerning nano-silica should not be generalized to all other types of nanomaterials. The environmental effects and overall outcomes when incorporating nanomaterials into asphalt depend on the manufacturing methods specific to each type of nanomaterial. Therefore, it is plausible that some nanomaterials could pose greater or minor environmental risks. Utilizing LCA for nanomaterial-modified asphalt mixtures offers a strategic approach for decision-makers to choose pavement modification additives, leveraging the advantages of nanomaterials in pavement construction while mitigating potential environmental hazards.

As emphasized throughout the document, comprehending the influence of data quality and methodological choices on LCA results is essential. These aspects significantly impact the reliability and comprehensiveness of LCA results, thus shaping the conclusions drawn and the decisions made based on them. Using high-quality data is essential in ensuring the credibility and validity of LCA results. Only after gaining a good understanding of these aspects, can we understand LCA results and decide on how to use them.

Along with the understanding of data quality and methodological choices, there is a need to understand the c-PCRs in different EPD programs. This enables meaningful comparisons between LCA results, particularly when presented in the form of EPDs for construction products within the same product category. The LCA results presented in the EPD should not merely be disseminated publicly but also leveraged to pinpoint areas where environmental impact reduction potential exists for the product. Furthermore, various standards within the EPD framework offer additional guidance on interpreting LCA results, enhancing their utility and applicability in decision-making processes.

In conclusion, while the utilization of the LCA methodology offers a wide range of advantages when applied to nanoproducts, there remains a considerable scope for further exploration in this domain. Future research is required by considering the analysis of the whole life cycle for nano-modified asphalt materials using different nanomaterials as a modifier to verify if nanomaterials are sustainable materials.

Bibliography

1. G. Ding, Life cycle assessment (LCA) of sustainable building materials: an overview. Life Cycle Assess. (LCA) Eco-Label. Case Stud., 38–62 (2014)
2. N. Nizam, M. Hanafiah, K. Woon, A content review of life cycle assessment of nanomaterials: current practices, challenges, and future prospects. Nanomaterials **11**(12), 3324 (2021)
3. I. Khan, K. Saeed, Nanoparticles: properties, applications and toxicities. Arab. J. Chem. **12**(7), 908–931 (2019)
4. G. Santhosh, G. Nayaka, Nanoparticles in construction industry and their toxicity, in *Ecological and Health Effects of Building Materials*, ed. by J. Malik, S. Marathe (Springer, 2022)
5. B. Salieri, D. Turner, B. Nowack, R. Hischier, Life cycle assessment of manufactured nanomaterials: where are we? NanoImpact **10**, 108–120 (2018)
6. K. Ettrup, A. Kounina, S. Foss Hansen, J.A.J. Meesters, E.B. Vea, A. Laurent, Correction to development of comparative toxicity potentials of tio2 nanoparticles for use in life cycle assessment. Environ. Sci. Technol. **51**(12), 7295 (2017)
7. REACHnano Consortium, Guidance for applying Life Cycle Assessment methodology to nanomaterials, 2015 [Online]. Available: https://invassat.gva.es/documents/161660384/162 311778/02+Guidance+for+applying+Life+Cycle+Assessment+methodology+to+nanoma terials/f40ee359-a279-4464-b2d0-c35c17d5922f. Accessed 25 Sept 2023
8. C. Dossche, V. Boel, W. De Corte, Use of life cycle assessments in the construction sector: critical review. Procedia Eng. **171**, 302–311 (2017)
9. European Commission, Climate Action: Progress Report 2022, December 2022 [Online]. Available: https://climate.ec.europa.eu/system/files/2022-12/com_2022_514_web_en.pdf. Accessed 03 Sept 2023
10. European Commission, Stepping up Europe's 2030 climate ambition Investing in a climate-neutral future for the benefit of our people, September 2020 [Online]. Available: https://eur-lex.europa.eu/legal-content/EN/TXT/?uri=CELEX%3A52020DC0562. Accessed 03 Sept 2023
11. P. Dräger, P. Letmathe, Value losses and environmental impacts in the construction industry—Tradeoffs or correlates? J. Clean. Prod. **336**, 130435 (2022)
12. R. Spence, H. Mulligan, Sustainable development and the construction industry. Habitat Int. **19**(3), 279–292 (1995)
13. UN, 2022 Global Status Report for Buildings and Construction, February 2023 [Online]. Available: https://globalabc.org/sites/default/files/2023-03/2022%20Global%20Status%20R eport%20for%20Buildings%20and%20Construction_1.pdf. Accessed 08 Sept 2023

© The Editor(s) (if applicable) and The Author(s) 2025
P. Mercader-Moyano and P. Porras-Pereira (eds.), *Life Cycle Analysis Based on Nanoparticles Applied to the Construction Industry*,
https://doi.org/10.1007/978-3-031-79115-4

14. IPCC Report, Climate Change 2021: The Physical Science Basis, the Working Group I Contribution to the Sixth Assessment Report, July 2022 [Online]. Available: https://www.ipcc.ch/report/sixth-assessment-report-working-group-i/. Accessed 08 Sept 2023

15. I. Ryłko-Polak, W. Komala, A. Białowiec, The reuse of biomass and industrial waste in biocomposite construction materials for decreasing natural resource use and mitigating the environmental impact of the construction industry: a review. Materials **15**(12), 4078 (2022)

16. M. Khasreen, P. Banfill, G. Menzies, Life-cycle assessment and the environmental impact of buildings: a review. Sustainability **1**(3), 674–701 (2009)

17. A.J. Probert, A. Miller, K. Ip, K.P. Beckett, R. Schofield, Accounting for the life-cycle carbon emissions of new dwellings in the U.K, in *12th International Conference on Non-conventional Materials and Technologies* (2010) pp. 1–13

18. I. Vasilca, M. Nen, O. Chivu, V. Radu, C. Simion, N. Marinescu, The management of environmental resources in the construction sector: an empirical model. Energies **14**(9), 2489 (2021)

19. E. Açıkkalp, A. Hepbasli, C. Yucer, T. Karakoc, Advanced life cycle integrated exergoeconomic analysis of building heating systems: an application and proposing new indices. J. Clean. Prod. **195**, 851–860 (2018)

20. E. Bernal, D. Edgar, B. Burnes, Building sustainability on deep values through mindfulness nurturing. Ecol. Econ. **146**, 645–657 (2018)

21. L. Guardigli, M. Bragadin, F. Della Fornace, C. Mazzoli, D. Prati, Energy retrofit alternatives and cost-optimal analysis for large public housing stocks, Energy Build. **166**, 48–59 (2018)

22. B. Huang, X. Wang, H. Kua, Y. Geng, R. Bleischwitz, J. Ren, Construction and demolition waste management in China through the 3R principle. Resour. Conserv. Recycl. **129**, 36–44 (2018)

23. S.A. Miller, V. John, S. Pacca, A. Horvath, Carbon dioxide reduction potential in the global cement industry by 2050. Cem. Concr. Res. **114**, 115–124 (2018)

24. J. Fořt, R. Černý, Transition to circular economy in the construction industry: environmental aspects of waste brick recycling scenarios. Waste Manage. **118**, 510–520 (2020)

25. R. Agrawal, V. Wankhede, A. Kumar, A. Upadhyay, J. Garza-Reyes, Nexus of circular economy and sustainable business performance in the era of digitalization. Int. J. Product. Perform. Manag. **71**(3), 748–774 (2021)

26. M. Ghufran, K. Khan, F. Ullah, A. Nasir, A.Al Alahmadi, A. Alzaed, M. Alwetaishi, Circular economy in the construction industry: a step towards sustainable development, Buildings **12**(7), 1004 (2022)

27. UN, Report of the World Commission on Environment and Development: Our Common Future, 1987 [Online]. Available: http://www.un-documents.net/our-common-future.pdf. Accessed 08 Sept 2023

28. M. Zu Castell-Rüdenhausen, M. Wahlström, T. Fruergaard Astrup, C. Jensen, A. Oberender, P. Johansson, E. Waerner, Policies as drivers for circular economy in the construction sector in the Nordics, Sustainability **13**(16), 9350 (2021)

29. G. Benachio, M. Freitas, S. Tavares, Circular economy in the construction industry: a systematic literature review. J. Clean. Prod. **260**, 121046 (2020)

30. Y. Kalmykova, M. Sadagopan, L. Rosado, Circular economy—from review of theories and practices to development of implementation tools. Resour. Conserv. Recycl. **135**, 190–201 (2018)

31. F. Ullah, S. Sepasgozar, C. Wang, A systematic review of smart real estate technology: drivers of, and barriers to, the use of digital disruptive technologies and online platforms. Sustainability **10**(9), 3142 (2018)

32. C. Garcés-Ayerbe, P. Rivera-Torres, I. Suárez-Perales, D. Leyva-De La Hiz, Is it possible to change from a linear to a circular economy? An overview of opportunities and barriers for european small and medium-sized enterprise companies. Int. J. Environ. Res. Public Health **16**(5), 851 (2019)

33. T. Joensuu, H. Edelman, A. Saari, Circular economy practices in the built environment. J. Clean. Prod. **276**, 124215 (2020)

34. P. Ghisellini, C. Cialani, S. Ulgiati, A review on circular economy: the expected transition to a balanced interplay of environmental and economic systems. J. Clean. Prod. **114**, 11–32 (2016)
35. D. Masi, S. Day, J. Godsell, Supply chain configurations in the circular economy: a systematic literature review. Sustainability **9**(9), 1602 (2017)
36. K. Hobson, Trajectories in environmental politics, G. Hayes, S. Jinnah, P. Kashwan, D. M. Konisky, S. Macgregor, J. M. Meyer and A. R. Zito, Eds., London: Routledge, 2022, p. 324.
37. T. Hák, S. Janoušková, B. Moldan, Sustainable development goals: a need for relevant indicators. Ecol. Ind. **60**, 565–573 (2016)
38. A. Wieser, M. Scherz, S. Maier, A. Passer, H. Kreiner, Implementation of sustainable development goals in construction industry—a systemic consideration of synergies and trade-offs, in *IOP Conference Series: Earth and Environmental Science* (Graz, Austria, 2019)
39. B. Xia, A. Olanipekun, Q. Chen, L. Xie, Y. Liu, Conceptualising the state of the art of corporate social responsibility (CSR) in the construction industry and its nexus to sustainable development. J. Clean. Prod. **195**, 340–353 (2018)
40. P. Mansell, S. Philbin, E. Konstantinou, Delivering UN sustainable development goals' impact on infrastructure projects: an empirical study of senior executives in the UK construction sector. Sustainability **12**(19), 7998 (2020)
41. P. Howden-Chapman, J. Siri, E. Chisholm, R. Chapman, C. Doll, A. Capon, SDG 3: Ensure healthy lives and promote wellbeing for all at all ages, 2017 [Online]. Available: https://cou ncil.science/wp-content/uploads/2017/03/SDGs-interactions-3-healthy-lives.pdf. Accessed 08 Sept 2023
42. K. Chaudhary, P. Vrat, Circular economy model of gold recovery from cell phones using system dynamics approach: a case study of India. Environ. Dev. Sustain. **22**, 173–200 (2020)
43. S. Türkeli, R. Kemp, B. Huang, R. Bleischwitz, W. McDowall, Circular economy scientific knowledge in the European Union and China: a bibliometric, network and survey analysis (2006–2016). J. Clean. Prod. **197**(1), 1244–1261 (2018)
44. A. Wijewansha, G. Tennakoon, K. Waidyasekara, B. Ekanayake, Implementation of circular economy principles during pre-construction stage: the case of Sri Lanka. Built Environ. Proj. Asset Manag. **11**(4), 750–766 (2021)
45. M. Ormazabal, V. Prieto-Sandoval, R. Puga-Leal, C. Jaca, Circular economy in Spanish SMEs: challenges and opportunities. J. Clean. Prod. **185**, 157–167 (2018)
46. O. Hanseth, K. Lyytinen, Design theory for dynamic complexity in information infrastructures: the case of building internet. J. Inf. Technol. **25**(1), 1–19 (2010)
47. S. Trivedi, K. Snehal, B. Das, S. Barbhuiya, A comprehensive review towards sustainable approaches on the processing and treatment of construction and demolition waste. Constr. Build. Mater. **393**, 132125 (2023)
48. Z. Wu, A.T. Yu, L. Shen, G. Liu, Quantifying construction and demolition waste: an analytical review. Waste Manage. **34**(9), 1683–1692 (2014)
49. Official Journal of the European Union, 2014/955/EU: Commission Decision of 18 December 2014 amending Decision 2000/532/EC on the list of waste pursuant to Directive 2008/98/EC of the European Parliament and of the Council Text with EEA relevance, 30 December 2014 [Online]. Available: https://eur-lex.europa.eu/legal-content/EN/TXT/?uri=celex%3A3 2014D0955. Accessed 10 Sept 2023
50. Journal of Laws of 2021, Item 2151, "The Act of 17 November 2021 Amending the Act on Waste and Some Other Acts," [Online]. Available: https://isap.sejm.gov.pl/isap.nsf/DocDet ails.xsp?id=WDU20210002151. Accessed 10 Sept 2023
51. Regulation of the Minister of Environment, Natural Resources and Forestry of 24 December 1997 on the Classification of Waste, [Online]. Available: https://www.prawo.pl/akty/dz-u-1997-162-1135,16799265.html. Accessed 10 Sept 2023
52. Official Journal EUC: Luxembourg, Commission Notice on Technical Guidelines on the Classification of Waste (2018/C 124/01), 2018 [Online]. Accessed 10 Sept 2023
53. M. Norouzi, M. Chàfer, L.F. Cabeza, L. Jiménez, D. Boer, Circular economy in the building and construction sector: a scientific evolution analysis. J. Build. Eng. **44**, 102704 (2021)

54. M. Bravo, J. Brito, J. Pontes, L. Evangelista, Mechanical performance of concrete made with aggregates from construction and demolition waste recycling plants. J. Clean. Prod. **99**, 59–74 (2015)
55. United Nations Environment Programme, Annual Report 2021, 2022. [Online] Available: https://wedocs.unep.org/bitstream/handle/20.500.11822/37946/UNEP_AR2021. pdf. Accessed 05 Sept 2023
56. J. Rodrigue, C. Comtois, B. Slack, *The Geography of Transport Systems* (Routledge, London, 2016)
57. J. Liu, H. Mooney, V. Hull, S.J Davis, J Gaskell, T. Hertel, J. Lubchenco, K. Seto, P. Gleick, C. Kremen, S. Li Systems integration for global sustainability. Science **347**, 6225 (2015)
58. L. Wang, X. Xue, Z. Zhao, Z. Wang, The impacts of transportation infrastructure on sustainable development: emerging trends and challenges. Int. J. Environ. Res. Public Health **15**(6), 1172 (2018)
59. D. Kammen, D.A. Sunter, City-integrated renewable energy for urban sustainability. Science **352**(6288), 922–928 (2016)
60. L. Kraus, H. Proff, Sustainable urban transportation criteria and measurement—a systematic literature review. Sustainability **13**(13), 7113 (2021)
61. United Nations, "Sustainable transport, sustainable development. Interagency report for second Global Sustainable Transport Conference.," 2021 [Online]. Available: https://sdgs. un.org/sites/default/files/2021-10/Transportation%20Report%202021_FullReport_Digital. pdf. Accessed 05 Sept 2023
62. European Environment Agency, "Transport and environment report 2021. Decarbonising road transport—the role of vehicles, fuels and transport demand," 2022 [Online]. Available: https:// www.eea.europa.eu/publications/transport-and-environment-report-2021. Accessed 05 Sept 2023
63. European Environment Agency, "Transport and environment report 2022. Digitalisation in the mobility system: challenges and opportunities," 2022 [Online]. Available: https://www.eea.europa.eu/publications/transport-and-environment-report-2022/transp ort-and-environment-report/view. Accessed 05 Sept 2023
64. C. Jeon, A. Amekudzi-Kennedy, R. Guensler, Evaluating plan alternatives for transportation system sustainability: atlanta metropolitan region. Int. J. Sustain. Transp. **4**(4), 227–247 (2010)
65. G. Ševčenko-Kozlovska, K. Čižiūnienė, The impact of economic sustainability in the transport sector on GDP of neighbouring countries: following the example of the baltic states. Sustainability **14**(6), 3326 (2022)
66. M. Gómez Garzón, "Nanomateriales, nanopartículas y síntesis verde," Revista Repertorio de Medicina y Cirugía. Fundación Universitaria de Ciencias de la Salud—FUCS, 2018
67. W. Shin, J. Cho, A.G. Kannan, Y. Lee, D. Kim, Cross-linked composite gel polymer electrolyte using mesoporous methacrylate-functionalized SiO_2 nanoparticles for Lithium-Ion polymer batteries. Sci. Rep. **6**, 26332 (2016)
68. A. Brinch, S. Hansen, N. Hartmann, A. Baun, EU regulation of nanobiocides: challenges in implementing the biocidal product regulation (BPR). Nanomaterials **6**(2), 33 (2016)
69. V. Makarov, A. Love, O. Sinitsyna, S. Makarova, I. Yaminsky, M. Taliansky, N. Kalinina, Green nanotechnologies: synthesis of metal nanoparticles using plants **6**(1), 35–44 (2014)
70. A. Boroumand Moghaddam, F. Namvar, M. Moniri, P.M. Tahir, S. Azizi, R. Mohamad, Nanoparticles biosynthesized by Fungi and Yeast: a review of their preparation, properties, and medical applications. Molecules, **20**(9), 16540–16565 (2015)
71. M. Buyle, J. Braet, A. Audenaert, Life cycle assessment in the construction sector: a review. Renew. Sustain. Energy Rev. **26**, 379–388 (2013)
72. I. Zabalza Bribián, A. Valero Capilla, A. Aranda Usón, Life cycle assessment of building materials: comparative analysis of energy and environmental impacts and evaluation of the eco-efficiency improvement potential. Build. Environ. **46**(5), 1133–1140 (2011)
73. K. Yahya, H. Boussabaine, A. Alzaed, Using life cycle assessment for estimating environmental impacts and eco-costs from the metal waste in the construction industry. Manag. Environ. Qual. **27**(2), 227–244 (2016)

74. Ecochain, "Life Cycle Assessment (LCA)—Complete Beginner's Guide," [Online]. Available: https://ecochain.com/blog/life-cycle-assessment-lca-guide/. Accessed 06 Sept 2023
75. Sustainable Facilities Tool, "Life Cycle Assessment—GSA Sustainable Facilities Tool," [Online]. Available: https://sftool.gov/plan/400/life-cycle-assessment. Accessed 05 Sept 2023
76. United Nations Environment Programme, "What is Life Cycle Thinking?" [Online]. Available: https://www.lifecycleinitiative.org/activities/what-is-life-cycle-thinking/. Accessed 06 Sept 2023
77. Y. Pamu, V. Kumar, M. Shakir, H. Ubbana, Life Cycle Assessment of a building using open-LCA software. Mater. Today: Proc. **52**(3), 1968–1978 (2022)
78. S. Kotaji, A. Schuurmans, S. Edwards (eds) *Life-cycle assessment in building and construction: a state-of-the-art report Society of Environmental Toxicology and Chemistry* (SETAC) (2003)
79. I. Blom, L. Itard, A. Meijer, Environmental impact of building-related and user-related energy consumption in dwellings. Build. Environ. **46**(8), 1657–1669 (2011)
80. L.A. Greening, D.L. Greene, C. Difiglio, Energy efficiency and consumption—the rebound effect—a survey. Energy Policy **28**(6–7), 389–401 (2000)
81. O. Ortiz-Rodríguez, F. Castells, G. Sonnemann, Life cycle assessment of two dwellings: one in Spain, a developed country, and one in Colombia, a country under development. Sci. Total Environ. **408**(12), 2435–2443 (2010)
82. International Organization for Standardization (ISO), "ISO 14044:2006. Environmental management—life cycle assessment—requirements and guidelines," [Online]. Available: https://www.iso.org/obp/ui/#iso:std:iso:14044:ed-1:v1:en. Accessed 09 Sept 2023
83. Sustainable Facilities Tool, "LCA Key Terms," [Online]. Available: https://sftool.gov/plan/630/lca-key-terms. Accessed 09 Sept 2023
84. United Nations Environment Programme, "Glossary of Life Cycle Terms," [Online]. Available: https://www.lifecycleinitiative.org/activities/life-cycle-terminology-2/. Accessed 09 Sept 2023
85. G. Pallas, "PRé Sustainability. Life cycle-based sustainability standards and guidelines," 2021. [Online]. Available: https://pre-sustainability.com/articles/lca-standards-and-guidelines/. Accessed 09 Sept 2023
86. Ecochain, "Impact Categories (LCA)—Overview," [Online]. Available: https://ecochain.com/blog/impact-categories-lca/. Accessed 05 Sept 2023
87. G. Finnveden, A. Moberg, Environmental systems analysis tools—an overview. J. Clean. Prod. **13**(12), 1165–1173 (2005)
88. Ecochain, "LCI Databases in LCA: Function & Most-used databases," [Online]. Available: https://ecochain.com/blog/lci-databases-in-lca/. Accessed 05 Sept 2023
89. I. Zabalza Bribián, A. Aranda Usón, S. Scarpellini, Life cycle assessment in buildings: state-of-the-art and simplified LCA methodology as a complement for building certification. Build. Environ. **44**(12), 2510–2520 (2009)
90. B. López-Mesa, "Análisis de ciclo de vida de sistemas constructivos/Life cycle assessment of construction systems," in *Conference on Sustainability in Edification*, (2011)
91. A. Martínez-Rocamora, J. Solís-Guzmán, M. Marrero, LCA databases focused on construction materials: a review. Renew. Sustain. Energy Rev. **58**, 565–573 (2016)
92. ITeC, "BEDEC. Construction cost bases. Institute of construction technology of Catalonia ITeC," 2019 [Online]. Available: https://itec.es/banco-precios-bedec/. Accessed 06 Sept 2023
93. M. Marrero, C. Rivero-Camacho, M.D. Alba-Rodríguez, What are we discarding during the life cycle of a building? Case studies of social housing in Andalusia, Spain. Waste Manage. **102**, 391–403 (2020)
94. European Commission, "European Platform on Life Cycle Assessment," [Online]. Available: https://data.jrc.ec.europa.eu/collection/EPLCA. Accessed 20 May 2023
95. Ecoinvent Centre, "Ecoinvent," [Online]. Available: https://ecoinvent.org/database/. Accessed 27 May 2023
96. H.C.M. Althaus, Life cycle inventories of metals and methodological aspects of inventorying material resources in ecoinvent (7 pp). J. Life Cycle Assess. **10**, 43–49 (2005)

97. D. Kellenberger, H.J. Althaus, T. Künniger, M. Lehmann, "Life Cycle Inventories of Building Products. Final report ecoinvent Data v2.0 No. 7.," 2007 [Online]. Available: https://db.ecoinvent.org/reports/07_BuildingProducts.pdf?area=463ee7e58cbf8. Accessed 06 Sept 2023

98. R. Hischier, "Life Cycle Inventories of Packaging and Graphical Papers. Ecoinvent-Report No. 11," 2007 [Online]. Accessed 27 May 2023

99. Greenhouse Gas Protocol, "GaBi Database website. Construction Materials extension," [Online]. Available: https://ghgprotocol.org/gabi-databases. Accessed 27 May 2023

100. PlasticsEurope, "PlasticsEurope Eco-Profiles website," [Online]. Available: https://plasticseurope.org/sustainability/circularity/life-cycle-thinking/eco-profiles-set/. Accessed 20 May 2023

101. Athena SMI, "Athena Database Content," [Online]. Available: https://www.calculatelca.com/wp-content/uploads/2012/10/LCI_Databases_Products.pdf. Accessed 20 May 2023

102. NREL (National Renewable Energy Laboratory), "U.S. Life Cycle Inventory Database," [Online]. Available: https://www.nrel.gov/lci/. Accessed 20 May 2023

103. ADEME, "Base Carbone website," [Online]. Available: https://data.ademe.fr/datasets/base-carboner. Accessed 20 May 2023

104. ITeC, "BEDEC website," [Online]. Available: https://itec.es/servicios/bedec/. Accessed 20 May 2023

105. P. Mercader-Moyano, Cuantificación de los recursos consumidos y emisiones de CO2 producidas en las construcciones de Andalucía y sus implicaciones en el protocolo de Kioto, Sevilla, 2010

106. CPM, "CPM LCA Database website," [Online]. Available: http://cpmdatabase.cpm.chalmers.se/Start.asp. Accessed 28 May 2023

107. IINAS, "GEMIS—Global Emissions Model for integrated Systems website," [Online]. Available: http://www.iinas.org/gemis.html. Accessed 28 May 2023

108. ProBas, "ProBas database website," [Online]. Available: https://www.probas.umweltbundesamt.de/php/index.php. Accessed 28 May 2023

109. U. Hantsche, Abschätzung des kumulierten Energieaufwandes und der damit verbundenen Emissionen zur Herstellung ausgewählter Baumaterialien/Assessment of the cumulative energy consumption and related emissions for the production of selected construction materials, VDI-Berichte 1093, 1993

110. S. Vierra, Green Building Standards and Certification Systems, 2023 [Online]. Available: https://www.wbdg.org/resources/green-building-standards-and-certification-systems#rcas. Accessed 19 Sept 2023

111. J. Lee, S. Mahendra, P.J.J. Alvarez, Nanomaterials in the construction industry: a review of their applications and environmental health and safety considerations. ACS Nano $4(7)$, 3580–3590 (2010)

112. J. Kreuter, Nanoparticles—a historical perspective. Int. J. Pharm. $331(1)$, 1–10 (2007)

113. N. Martín, "Sobre Fullerenos, Nanotubos de Carbono y Grafenos," *Arbor,* vol 187, no. Extra_1, p. 115–131 (2011)

114. A. Valizadeh, H. Mikaeili, M. Samiei, S. Farkhani, N. Zarghami, M. Kouhi, A. Akbarzadeh, S. Davaran, Quantum dots: synthesis, bioapplications, and toxicity. Nano Rev. $7(480)$ (2012)

115. I. Seil, T. Webster, Antimicrobial applications of nanotechnology: methods and literature. Int. J. Nanomedicine 7, 2767–3278 (2012)

116. E. Taylor, T. Webster, Reducing infections through nanotechnology and nanoparticles. Int. J. Nanomedicine 6, 1463–1473 (2011)

117. F.J. Heiligtag, M. Niederberger, The fascinating world of nanoparticle research. MaterialsToday $16(7–8)$, 262–271 (2013)

118. W. Shin, J. Cho, A. Kannan, Y. Lee, D. Kim, Cross-linked Composite Gel Polymer Electrolyte using Mesoporous Methacrylate-Functionalized SiO_2 Nanoparticles for Lithium-Ion Polymer Batteries. Sci. Rep. $6(26332)$ 2016

119. A. Astefanei, O. Núñez, M. Galceran, Characterisation and determination of fullerenes: a critical review. Anal. Chim. Acta $882(2)$, 1–21 (2015)

120. E. Dreaden, A. Alkilany, X. Huang, C. Murphy, M. El-Sayed, The golden age: gold nanoparticles for biomedicine. Chem. Soc. Rev. **41**(7), 2740–2779 (2012)
121. S. Thomas, Harshita, P. Mishra, S. Talegaonkar, Ceramic nanoparticles: fabrication methods and applications in drug delivery. Curr. Pharm. Des. **21**(42), 6165–6188 (2015)
122. S. Ali, I. Khan, S. Khan, M. Sohail, R. Ahmed, A. Rehman, M. Shahid Ansari, M. Ali Morsy, Electrocatalytic performance of Ni@Pt core–shell nanoparticles supported on carbon nanotubes for methanol oxidation reaction. J. Electroanal. Chem. **795**, 17–25 (2017)
123. M. Mansha, I. Khan, N. Ullah, A. Qurashi, Synthesis, characterization and visible-light-driven photoelectrochemical hydrogen evolution reaction of carbazole-containing conjugated polymers. Int. J. Hydrogen Energy **42**(16), 10952–10961 (2017)
124. N.H. Abd Ellah, S.A. Abouelmagd, Surface functionalization of polymeric nanoparticles for tumor drug delivery: approaches and challenges. Expert. Opin. Drug Deliv. **14**(2), 201–214 (2017)
125. M. Gujrati, A. Malamas, T. Shin, E. Jin, Y. Sun, Z. Lu, Multifunctional cationic lipid-based nanoparticles facilitate endosomal escape and reduction-triggered cytosolic siRNA release. Mol. Pharm. **11**(8), 2734–2744 (2014)
126. Y.M. Lei, W.X. Huang, M. Zhao, Y.Q Chai, R Yuan, Y. Zhuo, Electrochemiluminescence resonance energy transfer system: mechanism and application in ratiometric aptasensor for Lead ion. Anal. Chem. **87**(15), 7787–7794 (2015)
127. S. Unser, I. Bruzas, J. He, L. Sagle, Localized surface plasmon resonance biosensing: current challenges and approaches. Sensors **15**(7), 15684–15716 (2015)
128. A. Kosmala, R. Wright, Q. Zhang, P. Kirby, Synthesis of silver nano particles and fabrication of aqueous Ag inks for inkjet printing. Mater Chem Phys **129**(3), 1075–1080 (2011)
129. M. Shaalan, M. Saleh, M. El-Mahdy, M. El-Matbouli, Recent progress in applications of nanoparticles in fish medicine: a review. Nanomedicine Nanotechnol Biol Med **12**(3), 701–710 (2016)
130. F. Ning, M. Shao, S. Xu, Y. Fu, R. Zhang, M. Wei, D. Evans, X. Duan, TiO$_2$/graphene/NiFe-layered double hydroxide nanorod array photoanodes for efficient photoelectrochemical water splitting. Energy Environ. Sci. **9**, 2633–2643 (2016)
131. V. Avasare, Z. Zhang, D. Avasare, I. Khan, A. Qurashi, Room-temperature synthesis of TiO$_2$ nanospheres and their solar driven photoelectrochemical hydrogen production. Energy Research **39**(12), 1714–1719 (2015)
132. M. Gawande, A. Goswami, F. Felpin, T. Asefa, X. Huang, R. Silva, X. Zou, R. Zboril, R. Varma, Cu and Cu-based nanoparticles: synthesis and applications in catalysis. Chem. Rev. **116**(6), 3722–3811 (2016)
133. A. Ali, H. Zafar, M. Zia, I. Ul Haq, A. Phull, J. Ali, A. Hussain, Synthesis, characterization, applications, and challenges of iron oxide nanoparticles. Nanotechnol. Sci. **9**, 49–67 (2016)
134. Y. Zhou, C.-K. Dong, L. Han, J. Yang, X. Du, Top-down preparation of active cobalt oxide catalyst. ACS Catal. **6**(10), 6699–6703 (2016)
135. Z. Wang, X. Pan, Y. He, Y. Hu, H. Gu, Y. Wang, Z. Wang, X. Pan, Y. He, Y. Hu, H. Gu, Y. Wang, Piezoelectric nanowires in energy harvesting applications. Adv. Mater. Sci. Eng. **2015**, 1–21 (2015)
136. M. Kot, T. Major, J. Lackner, K. Chronowska-Przywara, M. Janusz, W. Rakowski, Mechanical and tribological properties of carbon-based graded coatings. J. Nanomater. **2016**, 1–14 (2016)
137. European Union Observatory for Nanomaterials, "Regulation," [Online]. Available: https://euon.echa.europa.eu/regulation. Accessed 22 Sept 2023
138. European Chemicals Agency, Nanomaterials, [Online]. Available: https://echa.europa.eu/regulations/nanomaterials. Accessed 22 Sept 2023
139. A. Mohajerani, L. Burnett, J.V. Smith, H. Kurmus, J. Milas, A. Arulrajah, S. Horpibulsuk, A. Abdul Kadir, Nanoparticles in construction materials and other applications, and implications of nanoparticle use. Materials **12**(19), 3052 (2019)
140. H. Boostani, S. Modirrousta, Review of nanocoatings for building application. Procedia Engineering **145**, 1541–1548 (2016)

141. M. Kumar, M. Bansal, R. Garg, An overview of beneficiary aspects of zinc oxide nanoparticles on performance of cement composites. Mater. Today: Proc. **43**(2), 892–898 (2021)

142. A.M. Rashad, Effects of ZnO_2, ZrO_2, Cu_2O_3, CuO, $CaCO_3$, SF, FA, cement and geothermal silica waste nanoparticles on properties of cementitious materials—a short guide for Civil Engineer. Constr. Build. Mater. **48**, 1120–1133 (2013)

143. H. Tang, S. Li, Y. Zhang, Y. Na, C. Sun, D. Zhao, J. Liu, Z. Zhou, Radiative cooling performance and life-cycle assessment of a scalable MgO paint for building applications. J. Clean. Prod. **380**(2), 135035 (2022)

144. S. Mourdikoudis, R.M. Pallares, N.T.K. Than, Characterization techniques for nanoparticles: comparison and complementarity upon studying nanoparticle properties. Nanoscale **10**, 12871–12934 (2018)

145. Y. Wang, D. Wan, S. Xie, X. Xia, C.Z. Huang, Y. Xia, Synthesis of Silver Octahedra with controlled sizes and optical properties via seed-mediated growth. ACS Nano **7**(5), 4586–4594 (2013)

146. F. Kim, S. Connor, H. Song, T. Kuykendall, P. Yang, Platonic gold nanocrystals. Angew. Chem. **43**(28), 3673–3677 (2004)

147. A.J. Pugsley, C.L. Bull, A. Sella, G. Sankar, P.F. McMillan, XAS/EXAFS studies of Ge nanoparticles produced by reaction between Mg_2Ge and $GeCl_4$. J. Solid State Chem. **184**(9), 2345–2352 (2011)

148. C. Blanco-Andujar, Thesis: Sodium carbonate mediated synthesis of iron oxide nanoparticles to improve magnetic hyperthermia efficiency and induce apoptosis, 2014 [Online]. Available: https://www.researchgate.net/publication/311431656_Sodium_carbonate_mediated_synthe sis_of_iron_oxide_nanoparticles_to_improve_magnetic_hyperthermia_efficiency_and_ind uce_apoptosis. Accessed 24 September 2023

149. L. Le Trong, Thesis: Water-dispersible magnetic nanoparticles for biomedical applications : synthesis and characterisation, 2011 [Online]. Available: https://livrepository.liverpool.ac. uk/3062809/. Accessed 24 Sept 2023

150. E.-S.M. Durai, K.A. Abdullin, Ferromagnetic resonance of cobalt nanoparticles used as a catalyst for the carbon nanotubes synthesis. J. Magn. Magn. Mater. **321**(24), L69–L72 (2009)

151. Y.W. Jun, J.W. Seo, J. Cheon, Nanoscaling laws of magnetic nanoparticles and their applicabilities in biomedical sciences. Acc. Chem. Res. **41**(2), 179–189 (2008)

152. Y. Pan, S. Neuss, A. Leifert, M. Fischler, F. Wen, U. Simon, G. Schmid, W. Brandau, W. Jahnen-Dechent, Size-dependent cytotoxicity of gold nanoparticles. Small **3**(11), 1941–1949 (2007)

153. D.B. William, C.B. Carter, High-resolution TEM in transmission electron microscopy (Springer US, Boston, MA, 2009), p. 483–509

154. P.A. Buffat Mater, Electron diffraction and HRTEM studies of multiply-twinned structures and dynamical events in metal nanoparticles: facts and artefacts. Mater. Chem. Phys. **81**(2–3), 368–375 (2003)

155. B. Schaffer, W. Grogger, G. Kothleitner, F. Hofer, Comparison of EFTEM and STEM EELS plasmon imaging of gold nanoparticles in a monochromated TEM. Ultramicroscopy **110**(8), 1087–1093 (2010)

156. I. Maliszewska, Z. Sadowski, Synthesis and antibacterial activity of silver nanoparticles. J. Phys.: Conf. Ser. **146**(012024) (2009)

157. N.R. Panyala, E.M. Peña-Méndez, J. Havel, Silver or silver nanoparticles: a hazardous threat to the environment and human health? J. Appl. Biomed. **6**, 117–129 (2008)

158. J. Hou, L. Wang, C. Wang, S. Zhang, H. Liu, S. Li, X. Wang, Toxicity and mechanisms of action of Titanium dioxide nanoparticles in living organisms. J. Environ. Sci. **75**, 40–53 (2019)

159. V.K. Sharma, Aggregation and toxicity of Titanium dioxide nanoparticles in aquatic environment—a review. J. Environ. Sci. Health Tox Hazard Subst Environ Eng. **44**(14), 1485–1495 (2009)

160. C. Larue, J. Laurette, N. Herlin-Boime, H. Khodja, B. Fayard, A.M. Flank, F. Brisset, M. Carriere, Accumulation, translocation and impact of TiO_2 nanoparticles in wheat (Triticum aestivum spp.): influence of diameter and crystal phase. Sci. Total Environ. **431**, 197–208 (2012)

161. U. Song, H. Jun, B. Waldman, J. Roh, Y. Kim, J. Yi, E.J Lee, Functional analyses of nanoparticle toxicity: a comparative study of the effects of TiO_2 and Ag on tomatoes (Lycopersicon esculentum). Ecotoxicol. Environ. Saf. **93**, 60–67 (2013)
162. N. Dissanayake, K. Current, S. Obare, Mutagenic effects of Iron Oxide nanoparticles on biological cells. Int. J. Mol. Sci. **16**(10), 23482–23516 (2015)
163. V. Madhubala, T. Kalaivani, A. Kirubha, J.S. Prakash, V. Manigandan, H.K. Dara, Study of structural and magnetic properties of hydro/solvothermally synthesized α-Fe_2O_3 nanoparticles and its toxicity assessment in zebrafish embryos. Appl. Surf. Sci. **494**, 391–400 (2019)
164. X. Zhu, S. Tian, Z. Cai, Toxicity assessment of Iron Oxide nanoparticles in Zebrafish (Danio rerio) early life stages. PLoS ONE **7**(9), e46286 (2012)
165. N. Adam, F. Leroux, D. Knapen, S. Bals, R. Blust, The uptake and elimination of ZnO and CuO nanoparticles in Daphnia magna under chronic exposure scenarios. Water Res. **68**, 249–261 (2015)
166. L. Dai, G.T. Banta, H. Selck, V.E. Forbes, Influence of copper oxide nanoparticle form and shape on toxicity and bioaccumulation in the deposit feeder, Capitella teleta. Mar. Environ. Res. **111**, 99–106 (2015)
167. X. Gao, A. Avellan, S. Laughton, R. Vaidya, S.M. Rodrigues, E.A. Casman, G.V. Lowry, CuO nanoparticle dissolution and toxicity to wheat (Triticum aestivum) in Rhizosphere soil. Environ. Sci. Technol. **52**(5), 2888–2897 (2018)
168. F. Schwab, T. Bucheli, L. Lukhele, A. Magrez, B. Nowack, L. Sigg, K. Knauer, Are carbon nanotube effects on green algae caused by shading and agglomeration? Environ. Sci. Technol. **45**(14), 6136–6144 (2011)
169. K. Kwok, K. Leung, E. Flahaut, J. Cheng, S. Cheng, Chronic toxicity of double-walled carbon nanotubes to three marine organisms: influence of different dispersion methods. Nanomedicine **5**(6), 951–961 (2010)
170. X. Zhu, L. Zhu, Y. Chen, S. Tian, Acute toxicities of six manufactured nanomaterial suspensions to Daphnia magna. Nanoparticles Occup. Health **11**, 67–75 (2009)
171. P. Ghafari, C. St-Denis, M. Power, X. Jin, V. Tsou, H. Mandal, N. Bols, X. Shirley, Impact of carbon nanotubes on the ingestion and digestion of bacteria by ciliated protozoa. Nat. Nanotechnol. **3**(6), 347–351 (2008)
172. A. Trompeta, I. Preiss, F. Bem-Ami, Y. Benayahu, C. Charitidis, Toxicity testing of MWCNTs to aquatic organisms. RSC Adv. **9**(63), 36707–36716 (2019)
173. Y. Zhu, Y. Tang, Z. Ruan, Y. Dai, Z. Li, Z. Lin and H. Long, $Mg(OH)_2$ nanoparticles enhance the antibacterial activities of macrophages by activating the reactive oxygen species. J. Biomed. Mater. Res. Part A **109**(1) (2021)
174. M. Fu, J. Liang, S. Wang, C. Geng, W. Zhang, T. Wu, The response of microalgae chlorella sp. to free and immobilized ZrO_2 and $Mg(OH)_2$ nanoparticles: perspective from the growth characteristics. Environ. Eng. Sci. **37**(6) (2020)
175. A.M. Azzam, M.A. Shenashen, M.M. Selim, A.S. Alamoudi, S.A. El-Safty, Hexagonal $Mg(OH)_2$ Nanosheets as antibacterial agent for treating contaminated water sources. ChemistrySelect **2**(35), 11431–11437 (2017)
176. Instituto Nacional de Seguridad e Higiene en el Trabajo (INSHT), "Seguridad y salud en el trabajo con nanomateriales," 2015 [Online]. Available: https://www.insst.es/documents/94886/96076/sst+nanomateriales/bd21b71f-d5ec-4ee8-8129-a4fa58480968. Accessed 24 Sept 2023
177. C. Marambio-Jones, E. Hoek, A review of the antibacterial effects of silver nanomaterials and potential implications for human health and the environment. J. Nanopart. Res. **12**, 1531–1551 (2010)
178. T. Coccini, S. Grandi, D. Lonati, C. Locatelli, U. De Simone, Comparative cellular toxicity of titanium dioxide nanoparticles on human astrocyte and neuronal cells after acute and prolonged exposure. Neurotoxicology **48**, 77–89 (2015)
179. N.V. Zaitseva, M.A. Zemlyanova, M.S. Stepankov, A.M. Ignatova, Scientific prediction of magnesium oxide nanoparticles toxicity and assessment of its hazard for human health. Ekologiya Cheloveka (Human Ecology) **26**(2), 39–44 (2019)

180. A. Pietroiusti, H. Stockmann-Juvala, F. Lucaroni, K. Savolainen, Nanomaterial exposure, toxicity, and impact on human health, Wiley Interdiscip. Rev. Nanomed. Nanobiotechnol. **10**(5) (2018)

181. W. Nazaroff, Exploring the consequences of climate change for indoor air quality. Environ. Res. Lett. **8**(015022) (2013)

182. W. Jones, A. Gibb, C. Goodier, P. Bust, M. Song, J. Jin, Nanomaterials in construction—what is being used, and where? Constr. Mater. **172**(2), 49–62 (2019)

183. D. Jassby, T.Y. Cath, H. Buisson, The role of nanotechnology in industrial water treatment. Nat. Nanotechnol. **13**, 670–672 (2018)

184. A. Kunhikrishnan, H.K. Shon, N.S. Bolan, I. El Saliby, S. Vigneswaran, Sources, distribution, environmental fate, and ecological effects of nanomaterials in wastewater streams. Crit. Rev. Environ. Sci. Technol. **45**(4), 277–318 (2015)

185. Eurostat, Waste statistics, January 2023 [Online]. Available: https://ec.europa.eu/eurostat/sta tistics-explained/index.php?title=Waste_statistics#Waste_treatment. Accessed 22 Sept 2023

186. Organisation for Economic Cooperation and Development (OECD), Nanomaterials in waste streams: current knowledge on risks and impacts (OECD Publishing, Paris, 2016)

187. Y. Zheng, B. Nowack, Size-specific, dynamic, probabilistic material flow analysis of titanium dioxide releases into the environment. Environ. Sci. Technol. **55**(4), 2392–2402 (2021)

188. S. Rajkovic, N.A. Bornhöft, R. van der Weijden, B. Nowack, V. Adam, Dynamic probabilistic material flow analysis of engineered nanomaterials in European waste treatment systems. Waste Manage. **113**, 118–131 (2020)

189. K. Phalyvong, Y. Sivry, H. Pauwels, A. Gélabert, M. Tharaud, G. Wille and M. F. Benedetti, Occurrence and origins of cerium dioxide and titanium dioxide nanoparticles in the Loire river (France) by single particle ICP-MS and FEG-SEM Imaging. Front. Environ. Sci. **8** (2020)

190. K. Mehrabi, R. Kaegi, D. Günther, A. Gundlach-Graham, Emerging investigator series: auto-mated single-nanoparticle quantification and classification: a holistic study of particles into and out of wastewater treatment plants in Switzerland. Environ. Sci. Nano **8**, 1211–1225 (2021)

191. P. Hennebert, A. Anderson, P. Merdy, Mineral nanoparticles in waste: potential sources, occur-rence in some engineered nanomaterials leachates, municipal sewage sludges and municipal landfill sludges. J. Biotechnol. Biomater. **7**(261) (2017)

192. A. Gogos, J. Wielinski, A. Voegelin, H. Emerich, R. Kaegi, Transformation of cerium dioxide nanoparticles during sewage sludge incineration. Environ. Sci. Nano **6**, 1765–1776 (2019)

193. DaNa, Nanomaterials in waste, [Online]. Available: https://nanopartikel.info/en/basics/cross-cutting/nanomaterials-in-waste/. Accessed 23 Sept 2023

194. A. Boldrin, S. Hansen, A. Baun, N.I. Bloch Hartmann, T. Fruergaard Astrup, Environ-mental exposure assessment framework for nanoparticles in solid waste. J. Nanoparticle Res. **16**(2394) (2014)

195. L. Heggelund, S. Foss Hansen, T. Fruergaard Astrup, A. Boldrin, Semi-quantitative analysis of solid waste flows from nano-enabled consumer products in Europe, Denmark and the United Kingdom—Abundance, distribution and management. Waste Manag. **56**, 584–592 (2016)

196. Austrian Academy of Sciences, NanoTrust-Dossier No. 040en: "Nanowaste" Nanomaterial-containing products at the end of their life cycle, 2014 [Online]. Available: https://epub.oeaw.ac.at/0xc1aa500e_0x003146a3.pdf

197. A. Caballero-Guzman, T. Sun, B. Nowack, Flows of engineered nanomaterials through the recycling process in Switzerland. Waste Manage. **36**, 33–43 (2015)

198. A. Keller, S. McFerran, A. Lazareva, S. Suh, Global life cycle releases of engineered nanomaterials. J. Nanoparticle Res. **15**(1692) (2013)

199. A. Akhtar, A. Sarmah, Construction and demolition waste generation and properties of recycled aggregate concrete: a global perspective. J. Clean. Prod. **186**, 262–281 (2018)

200. J. Gálvez-Martos, D. Styles, H. Schoenberger, B. Zeschmar-Lahl, Construction and demolition waste best management practice in Europe. Resour. Conserv. Recycl. **136**, 166–178 (2018)

201. Q. Chen, Q. Zhang, C. Qi, A. Fourie, C. Xiao, Recycling phosphogypsum and construction demolition waste for cemented paste backfill and its environmental impact. J. Clean. Prod. **186**, 418–429 (2018)

202. J. Tavira, J. Jiménez, J. Ayuso, A. López-Uceda, E. Ledesma, Recycling screening waste and recycled mixed aggregates from construction and demolition waste in paved bike lanes. J. Clean. Prod. **190**, 211–220 (2018)
203. M. Baalousha, Y. Yang, M. Vance, B. Colman, S. McNeal, J. Xu, J. Blaszczak, M. Steele, E. Bernhardt, M. Hochella, Outdoor urban nanomaterials: the emergence of a new, integrated, and critical field of study. Sci. Total Environ. **557–558**, 740–753 (2016)
204. M. Marcoux, M. Matias, F. Olivier, G. Keck, Review and prospect of emerging contaminants in waste—key issues and challenges linked to their presence in waste treatment schemes: general aspects and focus on nanoparticles. Waste Manage. **33**(11), 2147–2156 (2013)
205. M. Ferreira, E. Soldado, G. Borsoi, M. Mendes, I. Flores-Colen, Nanomaterials applied in the construction sector: environmental, human health, and economic indicators. Appl. Sci. **13**(23) (2023)
206. H. Babaizadeh, M. Hassan, Life cycle assessment of nano-sized titanium dioxide coating on residential windows. Constr. Build. Mater. **40**, 314–321 (2013)
207. F. Piccinno, F. Gottschalk, S. Seeger, B. Nowack, Industrial production quantities and uses of ten engineered nanomaterials in Europe and the world. J. Nanoparticle Res. **14**(1109) (2012)
208. A. Weir, P. Westerhoff, L. Fabricius, K. Hristovski, N. Goetz, Titanium dioxide nanoparticles in food and personal care products. Environ. Sci. Technol. **46**(4), 2242–2250 (2012)
209. T. Meng, H. Yu, Y. Zhao, S. Ruan, Strength/Service life-oriented environmental assessment of normal and recycled concrete with nano-SiO_2 in a natural and a marine environment: a pioneer case study in Zhejiang Province. J. Build. Eng. **64**(105623) (2023)
210. G. Li, W. Hu, H. Cui, J. Zhou, Long-term effectiveness of carbonation resistance of concrete treated with nano-SiO_2 modified polymer coatings. Constr. Build. Mater. **201**, 623–630 (2019)
211. M. Zajac, J. Skibsted, P. Durdzinski, F. Bullerjahn, J. Skocek, M. Ben Haha, Kinetics of enforced carbonation of cement paste. Cem. Concr. Res. **131**(106013) (2020)
212. R. Grazuleviciene, R. Nadisauskiene, J. Buinauskiene, T. Grazulevicius, Effects of elevated levels of manganese and iron in drinking water on birth outcomes. Polish J. of Environ. Stud. **18**(5), 819–825 (2009)
213. M. Fakharpour, M. Gholizadeh Arashti, Effect of three MWCNTs-COOH (chiral, zigzag and armchair) on the mechanical properties and water absorption of cement-based composites. Fuller. Nanotub. Carbon Nanostructures **31**(5), 404–411 (2023)
214. O. Kharissova, L. Torres Martínez, B. Kharisov, Recent trends of reinforcement of cement with carbon nanotubes and fibers. Adv. Carbon Nanostructures 137–160 (2016)
215. A. Matanza, G. Vargas, I. Leon, M. Pousse, N. Salmon, C. Marieta, Life cycle analysis of standard and high-performance cements based on carbon nanotubes composites for construction applications, in *World Sustainable Building 2014 Conference*. (Barcelona, 2014)
216. M. Oliveira, M. Izquierdo, X. Querol, R. Lieberman, B. Saikia, L. Silva, Nanoparticles from construction wastes: a problem to health and the environment. J. Clean. Prod. **219**, 236–243 (2019)
217. M. Kim, D. Goerzen, P. Jena, E. Zeng, M. Pasquali, R. Meidl, D. Heller, Human and environmental safety of carbon nanotubes across their life cycle. Nat. Rev. Mater. **9**(1), 63–81 (2024)
218. N. Mueller, J. Buha, J. Wang, A. Ulrich, B. Nowack, Modeling the flows of engineered nanomaterials during waste handling. Environ. Sci. Process. Impacts **15**(1), 251–259 (2013)
219. T. Dutta, K. Kim, A. Deep, J.E. Szulejko, K. Vellingiri, S. Kumar, E.E. Kwon, S. Yun, Recovery of nanomaterials from battery and electronic wastes: a new paradigm of environmental waste management. Renew. Sustain. Energy Rev. **82**(3), 3694–3704 (2018)
220. S. Ali Mazari, E. Ali, R. Abro, F. Saleem Ahmed Khan, I. Ahmed, M. Ahmed, S. Nizamuddin, T. Hussain Siddiqui, N. Hossain, N. Mujawar Mubarak, A. Shah, Nanomaterials: Applications, waste-handling, environmental toxicities, and future challenges—a review. J. Environ. Chem. Eng. **9**(2), 105028 (2021)
221. B. Reck, T. Graedel, Challenges in metal recycling. Science **337**(6095), 690–695 (2012)
222. V. Kumar, A. Choudhary, P. Kumar, S. Sharma, Nanotechnology: Nanomedicine, Nanotoxicity and Future Challenges. Nanosci. Nanotechnol. Asia **9**(1), 64–78 (2019)

223. K. Soto, A. Carrasco, T. Powell, K. Garza, L. Murr, Comparative in vitro cytotoxicity assessment of some manufacturednanoparticulate materials characterized by transmissionelectron microscopy. J. Nanopart. Res. **7**(2–3), 145–169 (2005)
224. Z. Chen, S. Han, S. Zhou, H. Feng, Y. Liu, G. Jia, Review of health safety aspects of titanium dioxide nanoparticles in food application. NanoImpact **18**(100224) (2020)
225. A. Sukhanova, S. Bozrova, P. Sokolov, M. Berestovoy, A. Karaulov, I. Nabiev, Dependence of nanoparticle toxicity on their physical and chemical properties. Nano Rev. **13**(44) 2018
226. M. Akter, T. Sikder, M. Rahman, A. Ullah, K. Binte Hossain, S. Banik, T. Hosokawa, T. Saito, M. Kurasaki, A systematic review on silver nanoparticles-induced cytotoxicity: Physicochemical properties and perspectives. J. Adv. Res. **9**, 1–16 (2018)
227. X. Gao, G. Lowry, Progress towards standardized and validated characterizations for measuring physicochemical properties of manufactured nanomaterials relevant to nano health and safety risks. NanoImpact **9**, 14–30 (2018)
228. C. Carlson, S. Hussein, A. Schrand, L. Braydich-Stolle, K. Hess, R. Jones, J. Schlager, Unique cellular interaction of silver nanoparticles: size-dependent generation of reactive oxygen species. J. Phys. Chem. B **112**(43), 13608–13619 (2008)
229. E. Sidiropoulou, K. Feidantsis, S. Kalogiannis, G. Gallios, G. Kastrinaki, E. Papaioannou, M. Václavíková, M. Kaloyianni, Insights into the toxicity of iron oxides nanoparticles in land snails. Comp. Biochem. Physiol. C Toxicol. Pharmacol. **206–207**, 1–10 (2018)
230. A. Bekdemir, S. Liao, F. Stellacci, On the effect of ligand shell heterogeneity on nanoparticle/ protein binding thermodynamics. Colloids Surf. B Biointerfaces **174**, 367–373 (2019)
231. European Chemicals Agency, Study on the Product Lifecycles, Waste Recycling and the Circular Economy for Nanomaterials, November 2021 [Online]. Available: https:// euon.echa.europa.eu/documents/2435000/3268576/nano_lifecycles_euon_en.pdf/107f2bd6-8967-5466-8f48-5610d9120bbe?t=1636969415023. Accessed 22 Sept 2023
232. European Chemicals Agency, Risk assessment of nanomaterials—further considerations, 2017 [Online]. Available: https://euon.echa.europa.eu/documents/2435000/3268584/nano_ in_brief_en.pdf/295c5f46-0f1e-4ad5-72a5-81c44b45bdd5?t=1531477424187. Accessed 22 Sept 2023
233. European Commission Joint Research Centre Institute for Environment and Sustainability, International Reference Life Cycle Data System (ILCD) Handbook—General (2010)
234. M. Hauschild, M. Huijbregts, *Life Cycle Impact Assessment* (Springer, Dordrecht, 2015)
235. UNE Asociación Española de Normalización, UNE-EN 15804:2012+A2:2020/AC:2021. Sustainability of construction works—Environmental product declarations—Core rules for the product category of construction products, 2021 [Online]. Available: https://www.une. org/encuentra-tu-norma/busca-tu-norma/norma?c=norma-une-en-15804-2012-a2-2020-ac-2021-n0067399. Accessed 21 May 2024
236. C. Lee, F. Sánchez, G. Asbjörnsson, M. Evertsson, Application of production simulation combined with site-specific data to quantify the environmental performance of aggregate products at five pilot sites, in *Conference in Minerals Engineering* (Luleå, Sweden, 2024)
237. F. Konradsen, K. Hansen, A. Ghose, M. Pizzol, Same product, different score: how methodological differences affect EPD results. Int. J. Life Cycle Assess. **29**, 291–307 (2024)
238. The International EPD System, Global EPD programme for publication of ISO 14025 and EN 15804 compliant EPDs, [Online]. Available: https://www.environdec.com/about-us/the-international-epd-system-about-the-system. Accessed 14 May 2024
239. epd-norway, 2022 [Online]. Available: https://www.epd-norge.no/. Accessed 14 May 2024
240. EPD Denmark, About EPD Denmark, 2022 [Online]. Available: https://www.epddanmark. dk/om-epd-danmark/. Accessed 14 May 2024
241. EPD Ireland, Develop an EPD, 2023 [Online]. Available: https://www.igbc.ie/develop-an-epd/. Accessed 14 May 2024
242. B-EPD program operator, The Belgian EPD programme B-EPD, 2022 [Online]. Available: https://www.health.belgium.be/en/belgian-epd-programme-b-epd. Accessed 14 May 2024
243. M. Gelowitz, J. McArthur, Comparison of type III environmental product declarations for construction products: material sourcing and harmonization evaluation. J. Clean. Prod. **157**, 125–133 (2017)

244. S. Welling, S. Ryding, Distribution of environmental performance in life cycle assessments—
     implications for environmental benchmarking. Int. J. Life Cycle Assess. **26**, 275–289 (2021)
245. S. Solomon, L. Dong-Eun, K. Byung-Soo, Life cycle assessment for the production phase of
     nano-silica-modified asphalt mixtures. Appl. Sci. **9**(7), 1315 (2019)
246. F. Todescato, I. Fortunati, A. Minotto, R. Signorini, J. Jasieniak, R. Bozio, Engineering of
     semiconductor nanocrystals for light emitting applications. Materials **9**(8), 672 (2016)
247. Y. Dai, J. Zhao, X. Liu, X. Yu, Z. Jiang, Y. Bu, Z. Xu, Z. Wang, X. Zhu, B. Xing, Transformation
     and species identification of CuO nanoparticles in plant cells (Nicotiana tabacum). Environ.
     Sci. Nano **6**, 2724–2735 (2019)